Approximation Theory and Optimization
*Tributes to M. J. D. Powell*

# Approximation Theory and Optimization

*Tributes to M. J. D. Powell*

Edited by
M. D. Buhmann
A. Iserles

CAMBRIDGE UNIVERSITY PRESS
Cambridge, New York, Melbourne, Madrid, Cape Town, Singapore, São Paulo, Delhi

Cambridge University Press
The Edinburgh Building, Cambridge CB2 8RU, UK

Published in the United States of America by Cambridge University Press, New York

www.cambridge.org
Information on this title: www.cambridge.org/9780521118446



First published 1997
This digitally printed version 2009

*A catalogue record for this publication is available from the British Library*

ISBN 978-0-521-58190-5 hardback
ISBN 978-0-521-11844-6 paperback

# Contents

# Preface

This volume comprises ten invited papers, presented at the *Conference on Numerical Mathematics* in Cambridge, on 27–30 July 1996.

The occasion being the sixtieth birthday of Professor M.J.D. Powell FRS, the main themes of the Cambridge conference were optimisation and approximation theory. These are the two principal areas of Mike Powell's interest. During a distinguished career, first at AERE Harwell and, since 1976, as the John Humphrey Plummer Professor of Applied Numerical Analysis in Cambridge, Mike Powell made outstanding contributions to both subjects. It is impossible to imagine the present state of the art in either optimisation or approximation without Mike's contribution!

The birthday party was an excuse for all of us to celebrate numerical analysis, listen to a variety of outstanding talks, debate computational mathematics in convivial and stimulating surroundings and, needless to say, meet again many old friends. Although the emphasis was on themes in optimisation and approximation, the range of the talks spanned all of numerical mathematics.

The many contributions of Mike Powell to the state of the art in modern numerical analysis were the *leitmotif* of the conference. It is a measure of Mike's influence on the development of both subjects that his contributions are never far from the surface in these proceedings. The opening essay of the volume is devoted to a brief survey of Mike Powell's work and it is accompanied by a list of his publications.

Many invited talks at the conference aimed at presenting summaries of the state of the art in a particular subject, while others introduced new research results. They combine to give timeliness and lasting value to the book; their quality reflects the general high standard of presentations at the meeting, not just of the invited but also the submitted papers. Indeed, the introspective examination of Mike Powell's work during the last 35 years was an excellent opportunity to summarise the advances in nonlinear optimisation, spline theory, approximations with radial basis functions etc. We thank Mike for his many and profound contributions to numerical analysis and for giving us thereby the opportunity to hold this conference and collect the papers in this book. We are pleased to have been able to gather papers from almost all the invitees and thank them and all participants for their work.

Support for this conference came from the Department of Applied Mathematics and Theoretical Physics in Cambridge, from the London Mathematical Society and from Barrodale Computing Services. Thanks are also due to the staff of DAMTP and of Magdalene College, where many delegates lodged and where the 'birthday party' on 29 July took place. Particular gratitude is due to the younger members of the numerical analysis group in Cambridge, Aurelian Bejancu, Chris Faigle, George Goodsell, Sigitas Keras, Yunkang

Liu and Antonella Zanna, who have helped us both before and during the conference with a multitude of administrative arrangements. And we thank David Tranah and the staff of Cambridge University Press for their help with publishing this book.

We conclude on a personal note. The first of us was Mike Powell's student and the second his colleague in Cambridge for nineteen years. Our knowledge and enjoyment of mathematics have been immensely influenced and transformed by Mike and our lives enriched by his friendship. May we thus conclude by wishing Mike many fruitful and happy years in numerical mathematics! We raise our glasses to papers yet to come.

*Martin D. Buhmann* *Arieh Iserles*

# Submitted Talks

*"Visualizing the performance of minimization methods combined with deflation techniques"*
**G.S. Androulakis**, T.N. Grapsa and M.N. Vrahtis (University of Patras)

*"Shifted cardinal interpolation and elliptic functions"*
**B.J.C. Baxter** (Imperial College) and N. Sivakumar (Texas A&M University)

*"Fast evaluation of multiquadric and other radial basis functions"*
**R.K. Beatson** (University of Canterbury, New Zealand) and G.N. Newsam (Defence Science Organisation, Australia)

*"On the uniform and pointwise convergence of thin-plate splines"*
**A. Bejancu** (University of Cambridge)

*"Exponentially convergent linear rational interpolation between equidistant (and other) points?"*
**J.-P. Berrut** (University of Fribourg) and H.D. Mittelmann (Arizona State University)

*"Numerical solution of boundary integral equations by means of attenuation factors"*
J.-P. Berrut and **M. Reifenberg** (University of Fribourg)

*"Surface interpolation with constrained area"*
M. Bozzini and **M. Rossini** (University of Milan)

*"On the numerical solution of nonlinear Volterra integral equations with blow-up solutions"*
**H. Brunner** (Memorial University of Newfoundland)

*"Radial functions with compact support"*
**M.D. Buhmann** (ETH Zürich)

*"TV-norms and their applications"*
**O. Burdakov** and B. Merkulov (CERFACS, Toulouse)

*"Convexity preserving Powell–Sabin interpolants to scattered data"*
J.M. Carnicer and **M. Floater** (University of Saragossa)

*"Convergence of CG and GMRES in the Waveform Relaxation fixed point equation"*
**E. Celledoni** and S. Maset (University of Trieste)

*"Some indeterminate moment problems and Freud-like weights"*
Y. Chen and **M.E.H. Ismail** (University of South Florida)

*"Cubature formulas for two-dimensional integrals based on multivariate quasi-interpolating splines"*
C. Dagnino and **P. Lamberti** (University of Turin)

*"On the convergence of product formulas based on nodal spline interpolation for the numerical evaluation of certain two-dimensional Cauchy Principal Value integrals"*
C. Dagnino, **S. Perotto** and E. Santi (University of Turin)

*"FMG cycle with the preconditioning of explicit group-2 linear system"*
**D.S. Daoud** and D. Subasi (Eastern Mediterranean University, Turkey)

*"On theorems of the alternative and duality"*
**A. Dax** (Hydrological Service, Israel)

*"Piecewise monotonic divided differences for least squares smoothing of univariate data: A review of methods and software"*
**I.C. Demetriou** (University of Athens)

*"SQP interior-point methods for optimal control problems"*
J.E. Dennis (Rice University), **M. Heinkenschloss** (Virginia Polytechnic Institute) and L.N. Vicente (Rice University)

*"Peano kernels of noninteger order"*
**K. Diethelm** (University of Hildesheim)

*"Asymptotic properties of some generalized penalty methods for nonlinear programming"*
**C. Gonzaga** and R. Castillo (Federal University of Santa Catarina, Brazil)

*"A new iterative method for thin plate spline interpolation to data points in general position"*
**G. Goodsell** (University of Cambridge)

*"Optimizing noisy functions using a one-dimensional bisection method"*
**T.N. Grapsa**, M.N. Vrahatis and G.S. Androulakis (University of Patras)

*"Problems and progress in vector Padé approximation"*
**P. Graves-Morris** (University of Bradford)

*"An inventory model with discrete transportation costs: An alternative approach for optimization"*
**P.N. Gupta** and V.K. Agarwal (JNV University, India)

*"On Lyapunov function method in Hilbert spaces"*
R.P. Ivanov (Bulgarian Academy of Sciences) and **I.L. Raykov** (Sheffield Hallam University)

*"A derivative-free algorithm for unconstrained minimization by searching conjugate directions on parallel hyperplanes"*
**H. Kanemitsu**, K. Yuriko, K. Hideaki and M. Shimbo (Hokkaido University)

*"Design of wavelets: A problem in optimization"*
**J. Kautsky** (Flinders University)

*"Combining domain decomposition with waveform relaxation"*
**S. Keras** (University of Bath)

*"Initial values for the inverse Toeplitz eigenproblem"*
**D. Laurie** (Potchefstroom University, South Africa)

*"Discrete approximation and convergence in $L^\infty$"*
**R. Lepp** (Institute of Cybernetics, Tallinn)

*"Analysis of a collocation method for integrating rapidly oscillatory functions"*
**D. Levin** (University of Tel Aviv)

*"Numerical methods for hybrid systems"*
**Y. Liu** (University of Cambridge)

*"Real versus interval methods for global optimization"*
**K. Madsen** (Technical University of Denmark)

*"Optimization problems arising in matrix and operator theory"*
**R. Mathias** (College of William and Mary)

*"Asymptotic error analysis in the finite element method"*
**D. Mayers** (University of Oxford)

*"The relationship between scaling and ill-conditioning in thin-plate spline systems"*
**G. Newsam** (Defence Science Organisation, Australia)

*"Numerical solution of partial differential equations with boundary singularities using the singularity subtraction technique"*
C.K. Pan and **K.M. Liu** (City University of Hong Kong)

*"Numerical estimation of projection constants"*
**K. Petras** and C. Helzel (Technical University of Braunschweig)

*"On the behaviour of a particular class of refinable functions"*
**F. Pitolli** (University of Rome La Sapienza)

*"Log-sigmoid rescaling in constrained optimization"*
**R.A. Polyak** (George Mason University)

*"Optimally choosing non-constant weighting for the smoothing spline"*
**B.C. Pringle** and K.W. Bosworth (Idaho State University)

*"Spectral properties of the modified Chebyshev pseudospectral method"*
**R. Renaut** (Arizona State University)

*"An algorithm for selecting a good value for the parameter c in radial basis interpolation methods"*
**S. Rippa** (Orbotech, Israel)

*"A new SQP algorithm for large-scale NLP"*
**R. Sargent** (Imperial College)

*"Extrapolation techniques and the collocation method for some boundary integral equations"*
**F.-J. Sayas** and R. Celorrio (University of Saragossa)

*"Powell–Sabin splines in restricted approximation"*
**J.W. Schmidt** (Technical University of Dresden)

*"Risk management and optimization"*
**D. Siegel** (IBM, Germany)

*"ABS linear solvers for optimization"*
**E. Spedicato** (University of Bergamo)

*"Fast Fourier transform on 2-spheres"*
**G. Steidl** (University of Darmstadt)

*"On approximate variants of classical iterative methods"*
**O. Vaarmann** (Tallinn Technical University)

*"Complex and analytic splines in the extremal problems of the approximation theory and the numerical analysis"*
**S.B. Vakarchuk** and V.I. Zabutnaya (National Ukrainian Academy of Sciences)

*"Spectral signature classification via wavelet coefficient nonparametric density estimation"*
**E. Velásquez** and K.W. Bosworth (Idaho State University)

*"Wavelet data denoising and function estimation: bayesian approach"*
**B. Vidakovic** (Duke University)

*"Modifications of Armijo's method"*
**M.N. Vrahatis**, G.S. Andrulakis and T.N. Grapsa (University of Patras)

*"A new efficent ODE solver of 3-stage, fourth order for solving initial value problems"*
N. Yaacob and **B.B. Sanugi** (Malaysian University of Technology)

*"Entropies of classical orthogonal polynomials with an unbounded supports"*
**R.J. Yáñez** (University of Granada), J.S. Dehesa (University of Granada), W. van Assche (Catholic University, Leuven) and A.I. Aptekarev (Moscow State University)

*"On the unicity of segmented $L_1$ approximations"*
**N. Yang** (Fen Jen Catholic University, Taiwan)

*"The method of iterated commutators for ordinary differential equations on Lie groups"*
**A. Zanna** (University of Cambridge)

*"Fitting B-spline curves and surfaces via ODR"*
**D. Zwick** (double star Research, St Augustin, Germany)

# M.J.D. Powell's Contributions to Numerical Mathematics

*Martin D. Buhmann*

Department of Mathematics, ETH Zentrum, 8092 Zürich, Switzerland

*Roger Fletcher*

Department of Mathematics and Computer Science, University of Dundee, Dundee DD1 4HN, UK

## 1 A brief review of M.J.D. Powell's work in univariate and multivariate approximation theory

Since 1966, about 30 years ago, Mike Powell has published more than 40 papers on approximation theory, initially mostly on univariate approximations and then, focussing especially on radial basis functions, on multivariate methods too. A highlight in his work is certainly his book "Approximation theory and methods", published by CUP in 1981, which summarizes and extends much of his work on $\ell_1$, $\ell_2$, $\ell_\infty$ theory and methods, splines, polynomial and rational approximation etc. It is still one of the best texts on univariate approximation theory available. In this short article we make an attempt to introduce part of Mike's work with special emphasis on splines in one dimension on one hand and radial basis functions on the other hand. Only a selection of his papers can be considered, and we will have to leave out all of his many software contributions which for Mike are an integral part of his research work, be it for the purpose of establishing new or better methods for approximation or for making them more accessible to the general public through library systems. We shall subdivide this section into three subsections ($\ell_1/\ell_\infty$ approximation, rational approximation; splines; multivariate (radial basis function) approximation) although this is really contrary to the spirit of many of Mike's articles which often establish beautiful links between various directions (e.g. optimization and $\ell_1$ approximation). As will be seen, many of the papers contain optimal results in the sense that constants in error estimates are best (or the best ones known), often have surprising novelties and always clearly defined goals. One further important contribution that we cannot describe here is Mike's guidance for the seven dissertations in approximation theory that were written under his supervision. We commence with his work on univariate approximation.

## 1.1 $\ell_1/\ell_\infty$ approximation, rational approximation

Mike's work on these topics begins with contributions to best polynomial approximation steps. A particular concern of his is their efficient computation, e.g. with the Remez (or exchange) algorithm. In his paper with Alan Curtis [13][1], the exchange algorithm for best $\ell_\infty$ approximation from finite dimensional (especially polynomial) spaces is studied with a view to its convergence properties. In particular, formulae for first and second partial derivatives of

$$\max_{\substack{x=x_i\\ i=0,1,\ldots,n}} |f(x) - \phi(x, \mu_1, \ldots, \mu_n)|$$

are given with respect to the reference points $x_i$, $i = 0, 1, \ldots, n$, where $f$ is to be approximated from $\{\phi(x, \mu_1, \ldots, \mu_n) | \mu_i \in \mathbb{R},\ i = 0, 1, \ldots, n\}$. If for instance the first and second derivatives are zero, the required stationary points are reached with quadratic convergence rate when the reference prints are moved one at a time. In other words, conditions are given in this paper under which the exchange algorithm is particularly efficient.

A year later, in the same journal, there appeared a paper [20] that considers error estimates of polynomial approximation. Lebesgue numbers (norm of the interpolation operator for polynomial interpolation) are used to show that the maximum error of interpolation is always (within a factor independent of $f$) a multiple of the least maximum error. Chebyshev points on an interval are demonstrated as a good choice of interpolation points. The same analysis is also applied to best least-squares approximation. In the following work [23], properties of optimal knot positions for the latter approximation method are studied.

Rational, rather than polynomial, $\ell_\infty$ approximation is the subject of a joint paper with Barrodale and Roberts [41] where it is shown that the ordinary differential correction (ODC) algorithm is, surprisingly, in various senses better than the modified one that is more often used. The ODC is employed to find a best $\ell_\infty$ approximation by $P/Q$, $P \in \mathbb{P}_m$, $Q \in \mathbb{P}_n$, on the basis of points $X = \{x_1, x_2, \ldots, x_N\}$, where $Q(x_i) > 0\ \forall i = 1, 2, \ldots, N$. Although the maximum error $\Delta_k$ over $X$ may tend to $\Delta^*$ in the course of the algorithm, where $P_k$, $Q_k$ are computed from $P_{k-1}$ and $Q_{k-1}$ by minimizing

$$\max_{i=1,2,\ldots,N} \frac{|f(x_i)\,Q(x_i) - P(x_i)| - \Delta_{k-1}\,Q(x_i)}{Q_{k-1}(x_i)},$$

the problem need not have a solution $P^*, Q^*$ with the required properties for which $\Delta^*$ is attained. However, convergence $P_k \to P^*$, $Q_k \to Q^*$ is proved in [41] when $N \geq n + m + 1$. The best approximation exists and satisfies the normalization condition $\max_i |q_i| = 1$. Here, the $q_i$ are the coefficients of $Q$. Furthermore, the minimization procedure involved is a linear programming

[1]Numbers in square brackets refer to the general list of M.J.D. Powell's publications.

problem, and quadratic convergence of the coefficients is established. Incidentally, this is a beautiful link between approximation theory and an application of an optimization (specifically, here, linear programming) method, four more of which we will encounter below.

Often, we find surprising novelties in his papers at points where nothing new was expected. Examples will be in his work on radial basis functions outlined below, but also the problem dealt with in the last paragraph was revisited for new results in [108]. There, general rational approximations are considered, i.e. ones where the numerator and the denominator need no longer be polynomial. However, it is sensible to restrict them to finite dimensional spaces, $G$ and $H$ say, and the denominator should be from the set $H_+ := \{h \in H \mid h > 0,\ \|h\| = 1\}$ which is assumed to be nonempty, and $\|\cdot\|$ is a prescribed norm.

Again, the unmodified, ordinary ODC algorithm is analyzed and superlinear convergence is shown if the best approximation with respect to $\|\cdot\|$ exists and is unique and if $\Delta_k \to \Delta^*$. If $\inf Q_n \to 0$ and no unique best approximation exists, an example of just linear convergence is given, even if $P_k$ and $Q_k$ tend to a unique limit for $k \to \infty$. Thus the former result is optimal with respect to its hypotheses. It is well known that the best approximation may not exist uniquely even if $\Delta_k \to \Delta^*$ but $\inf Q_k \to 0$, although $P_k$ and $Q_k$ may have a unique limit. Another approach to solve this problem, that has been suggested elsewhere, is to restrict $Q_k$ away from zero, $Q_k \geq \epsilon > 0$, where $\epsilon$ is independent of $k$, during the algorithm.

Another link between approximation and optimization occurs in $\ell_1$ theory. In [79], a linear programming test for $\ell_1$ optimality of an approximation is given. It consists of a finite number of linear inequalities and is therefore suitable for the application of an LP method. These inequalities are expressed in terms of the original data, which is highly suitable for practical computations of best $\ell_1$ approximations. Those approximations have been, incidentally, much neglected elsewhere in the literature, with the exception of A. Pinkus' excellent book on $\ell_1$ approximations. The characterization theorem is best stated explicitly. We let $X$ be a discrete finite set, $C(X)$ the set of functions that are defined on $X$, and $A \subset C(X)$ an $n$-dimensional subspace. Furthermore, we let $Z = \{z_j\}_{j=1}^n \subset X$ be the zero set with $n$ elements of the function $s(x) = f(x) - \phi^*(x)$, where $f \in C(X)$ is the function to be approximated, $\phi^* \in A$. Then $\phi^*$ is best $\ell_1$ approximant to $f$ if and only if the following inequalities hold:

$$\Big| \sum_{x \in X \setminus Z} \text{sign}\, s(x)\, \ell_i(x) \Big| \leq 1 \quad \forall i,$$

where $\ell_i \in A$ satisfy $\ell_i(z_j) = \delta_{ij}$, $i = 1, \ldots, n$. There are many advantages to using $\ell_1$ approximations in practice and this paper is a concrete aid to their application.

Finally in this subsection we mention the articles [128], [129], where approximants to discrete noisy data are constructed to obtain either piecewise monotonicity of the data or convexity. Precisely, if, for instance, at most $k$ monotone sections of the data are desired ($k$ is prescribed) and $n$ (univariate) data are given, the $k$ optimal breakpoints and the least changes to the data are computed in $O(n^2 + kn\log n)$ operations. "Least changes" is understood in the sense of a global sum of squares of changes. The special cases $k = 1, 2$ give the minimum complexity $O(n)$. The principal advancement in this work is the substantial reduction of the number of data that need be considered when finding the optimal breakpoints. A recursive method is applied with respect to $k$, and certain subsets of the data have to be used in the computation at each stage.

In [129], the least sum of squares of changes to the data is sought to achieve convexity. The method uses the iterative optimization algorithm of Goldfarb and Idnani, for which a starting value is computed in $O(n)$ operations. Precisely, the statement of the convexity constraints in terms of second divided differences of the data gives rise to a strictly convex quadratic programming problem which is subsequently solved by the above algorithm. Mike's talk at the conference celebrating his 60$^{\text{th}}$ birthday this year was also closely related to this topic. In [89], similar questions are considered with respect to the least uniform change to the data. Algorithms are given to compute the least maximum change to the data in order to achieve monotonicity, piecewise monotonicity, or piecewise convex/concave data.

## 1.2 Splines

In this subsection, a short summary of some of Mike's work on splines is presented. The best source for this work is generally his book [81] where much attention is given to polynomial splines, B-splines, spline interpolation and convergence properties, but we extract his work here from the papers [29], [32] and [63].

Cubic splines are considered in [29] and they are employed for the purpose of providing least squares approximations with weights. The main purpose of the article is an analysis of the locality of the least squares approximation by splines. The (finite number of) knots $\ell h, \ell \in \mathbb{Z} \cap [0, M]$, say, of the spline are equidistant with spacing $h$ and the weight function is $h$-periodic and nonnegative. Using recurrence relations, the fundamental functions for this spline approximation problem are computed where, for simplicity, the knots are assumed to be all $\ell h$, $\ell \in \mathbb{Z}$. They decay exponentially unless the weight is concentrated solely at the midpoints between the knots by a $\delta$-function $h \cdot \delta(x - \frac{1}{2}h)$ (precisely, by its $h$-periodization). The fastest decay of $(2-\sqrt{3})^{\ell}$, $\ell \to \pm\infty$, is obtained when the weight is concentrated at the knots, by $h \cdot \delta(x)$. A further consideration is given to the case when the weighted $\ell_2$-norm is augmented by the sum of squares of the coefficients $c_j$ where the spline is $s(x) = \sum_{j=0}^{M} c_j (x - jh)_+^3$ plus a cubic polynomial, and this augmentation may again be weighted by a positive factor $\vartheta$. Indeed, by employing this factor, the localization of the spline's dependence on the data can be strengthened, and also the best $\vartheta$ is given for the most unfavourable choice of the weight function in the least squares integral, namely $h \cdot \delta(x - \frac{1}{2}h)$, when the knots are still presumed to be integer multiples of $h > 0$. The fundamental function centred at 0 for the best approximation using that weight function and $\vartheta$ is shown to diminish as $(0.3613)^{\ell}$, when its argument is $\ell h$, $\ell \in \mathbb{Z}$. This damping term can also give uniqueness of the best approximant, which may otherwise be lost when the weight function is not periodic. Moreover, differences of the approximant are considered, e.g. if the eighth derivative of the approximand is bounded, they are $O(h^8)$ and if $f \in \mathbb{P}_7$, the error is zero except for dependencies at the ends of the range, when the approximand and approximant are defined on a finite interval.

In [27], the norm of the spline interpolation operator is estimated as well as the deterioration of the localization of the fundamental functions when the degree of the splines becomes larger. In [32], least squares approximations to discrete data, and in particular an adaptive method for computing them when the knots are allowed to move, are considered and we describe the approach in some detail now. These papers reflect Mike's interest in smoothing techniques, just like the articles discussed at the end of the previous subsection, which are highly relevant to many applications.

In [32], as in [29], a weighted sum of squares of the discrete error plus a smoothing term is to be minimized by a cubic spline, whose knots are to be determined. The smoothing term is itself a weighted sum of squares, namely of the coefficients of the truncated third powers that appear in the spline; those coefficients reflect the contribution of the third derivative discontinuities of the spline and their size is therefore a measure of its smoothness. Mike makes a distinction between knots and gnots of the spline in this paper, the latter being added whenever there is a trend found in the residual (the error function at the data ordinates). The test for trends is applied locally between the current gnots. The former knots are added so that the total distribution of the splines breakpoints remains balanced in a certain sense even if gnots are accumulating. Particular attention is given to the weights in the smoothing expression which depend on the distance of knots and gnots and on the weighted sum of squares of the residuals. We recall from [29] that the smoothing term can also cause the approximation to depend more locally on the data.

Another view of optimal knot positions, now for plain, unweighted least-squares approximation, is presented in the theoretical work [23]. Splines of degree $n$ with $N$ knots are studied and the goal is to minimize

$$\int_a^b (f(x) - s(x))^2 \, dx \, ,$$

where the function $f \in L^2[a, b]$ is bounded. A necessary condition for the optimality of a knot $x_j$ is

$$\int_a^b f(x)\, \tilde{c}_j^{(n+1)}(x)\, dx = 0\,,$$

where $\tilde{c}_j$ is the spline of degree $2n+1$ with simple knots $a < x_1 < x_2 < \ldots < x_N < b$ and an extra knot at $x_j$, which satisfies the Hermite conditions

$$\begin{aligned} \tilde{c}_j^{(p)}(a) &= \tilde{c}_j^{(p)}(b)\,, \quad 0 \le p \le n\,, \\ \tilde{c}_j(x_\ell) &= 0\,, \qquad\quad 1 \le \ell \le N\,, \\ \tilde{c}_j'(x_j) &= 1\,. \end{aligned}$$

The above condition is a consequence of the orthogonality of $\tilde{c}_j^{(n+1)}$ to all splines with knots $x_1 < x_2 < \ldots < x_N$ and an explicit expression for the $n$-fold integral of the approximation's error function evaluated at $x_j$. A suitable minimization algorithm is available from [8] for minimizing the error functional with respect to the varying knots. An essential tool in deriving the above condition is the application of integration by parts in order to reformulate orthogonality relations with splines to point evaluations of splines of higher degree. The article [23] is one of the first papers where this highly useful technique is used.

In the 1970s, the optimal interpolation problem was widely discussed. In particular, much attention was paid to finding the least pointwise error bound using a multiple (that is to be determined pointwise) of $\|f^{(k)}\|_\infty$, $k$ prescribed. Here $f$ is the function to be approximated, of sufficient smoothness, and $\|\cdot\|_\infty$ is the uniform norm over the interval where $f$ is defined. Hence an approximant $s$ and a function $c$ are sought, where $c$ is smallest for every $x$ while satisfying

$$|f(x) - s(x)| \le c(x)\, \|f^{(k)}\|_\infty\,,$$

and $s$ should be a linear combination of the given values of $f$ at $\{x_i\}_{i=1}^m$, $m \ge k$, in the interval; the coefficients depending on $x$ of course. It turns out that $s$ is a spline of degree $k-1$ with $m-k$ knots, and so is $c$. This is identified in [63] jointly with P. Gaffney by solving the problem first for $\|f^{(k)}\|_\infty = M$, i.e. a fixed value, by

$$\begin{aligned} s = s_M &= \frac{1}{2}\,(u(x,M) + \ell(x,M)) \\ c = c_M &= \frac{1}{2M}\,|u(x,M) - \ell(x,M)|. \end{aligned}$$

Here $u$ and $\ell$ are defined by

$$\frac{\partial^k}{\partial x^k}\, u = \delta_u(x), \qquad \frac{\partial^k}{\partial x^k}\, \ell = \delta_\ell(x)$$

and $\delta_\ell$, $\delta_u$ are the piecewise constant functions with $|\delta_u| = |\delta_\ell| = \|\delta_u\|_\infty = \|\delta_\ell\|_\infty = M$ and have $m-k$ sign changes consistent with the signs of the data. Their signs alternate, beginning with $+1$ for $\delta_u$ and $-1$ for $\delta_\ell$, and their discontinuity points have to be computed by solving a certain nonlinear system of equations. The final $s$ and $c$ are thus obtained by letting $M \to \infty$.

In [68] a review of bivariate approximation tools is given, mostly tensor product methods, according to the state of the art at the time, where not just point interpolation, but general linear operators are considered; e.g. least-squares approximation is considered. Also irregular data receive attention; Mike especially discusses weighted local least square approximation by piecewise linears, considers triangulations of domains in two dimensions, etc. A new state-of-the-art conference has taken place this year (1996) in York, and Mike spoke again on multivariate approximation. It is interesting to compare the two papers – twenty years apart – and observe the development of the subject. Indeed, in the new paper there is much more attention given to general multivariate tools, albeit only for interpolation. Most of the work reviewed concerns radial functions which we describe in the following section. Piecewise linears and higher order polynomials are discussed in connection with algorithms for generating Delauney triangulations. Local interpolation schemes, such as Shephard's method and its generalizations are mentioned too, as are moving least squares and natural neighbourhood interpolants.

The article [69] deserves an especially prominent mention because the Powell–Sabin split is very well-known and much used in the CAGD and finite element communities. The theme of that paper is the creation of a globally continuously differentiable piecewise quadratic surface on a triangulation. Function values and derivatives are given at the vertices of each triangle of a partition. In order to have the necessary number of degrees of freedom, each triangle is subdivided into six (or twelve) subtriangles which require additional, interior $C^1$ conditions. In total, the number of degrees of freedom turns out to be precisely right (nine) in the six subtriangle case; extra degrees of freedom can be taken up by prescribing also normals on edges which may be computed by linear interpolation in an approximation. How are the interior subtriangles selected? In the six subtriangle case, that is the one most often referred to as the Powell–Sabin element, one takes the edges of the interior triangles from the midpoints of the edges of the big triangle to its circumcentre. The triangles are required to be always acute. In other words, the midpoint inside the big triangle is the intersection of the normals at the midpoints of the edges. This ensures that the midpoint inside the big triangle lies in the plane spanned by the points exactly between it and the edge's midpoint, which is needed for the interior $C^1$ continuity. As indicated above, extra degrees of freedom are there when the twelve triangle split is used and thus, e.g., the condition that all triangles of the triangulation be acute may be dropped. This paper is just another case where Mike's work (and, of course, that of his co-authors, although most of his papers he has written alone) initiated a large amount of lasting interest in research and applications. Another example displaying much foresight is his work on radial basis functions that we will describe in the next section.

## 1.3 Radial basis functions

Mike Powell was and still is one of the main driving forces behind the research into radial basis functions. It indeed turned out to be a most successful and fruitful area, leading to many advances in theory and applications. Much of the interest within the mathematical community was stimulated by Mike's first review paper [107]; which discusses recent developments in this new field. It addresses the nonsingularity properties of the interpolation matrices $\{\phi(\|x_i - x_j\|_2)\}_{i,j=1}^m$ for interpolation at distinct points $x_j \in \mathbb{R}^n$, where $\phi(r) = r$ or $\phi(r) = e^{-r^2}$ or $\phi(r) = \sqrt{r^2 + c^2}$, $c$ a positive parameter. Indeed, for any $n$ and $m$, those matrices are always nonsingular, admitting unique solvability of the pointwise interpolation problem from the linear space spanned by $\phi(\| \cdot - x_j\|_2)$. Proofs of these results are provided and examples of singularity are given for 1-norms and $\infty$-norms replacing the Euclidean norm. It is incidentally indicated in that paper that Mike's interest in radial basis function methods was stimulated by the possibility of using them to provide local approximations to functions required within optimization algorithms.

Motivated by these remarkable nonsingularity results, at first two of Mike's research students worked in this field, their early results being summarized in [114]. Their work concentrated initially on the question whether polynomials are contained in the linear spaces spanned by (multi-) integer translates of the radial basis function $\phi(r) = \sqrt{r^2 + c^2}$, where the underlying space is $\mathbb{R}^n$. For odd $n \geq 1$, all polynomials of degree $n$ were shown to exist in those spaces for $c = 0$. For $n = 1$, $c > 0$, the same was shown for linear polynomials. In both cases, the polynomials $p$ are generated by quasi-interpolation

$$p(x) = \sum_{j \in \mathbb{Z}^n} p(j)\, \psi(x - j)\,, \quad x \in \mathbb{R}^n\,,$$

where $\psi$ is a finite linear combination of integer translates of those radial basis functions. The first task is to show that $\psi$ exist which decay sufficiently fast to render the above infinite sum absolutely convergent when $p$ is a polynomial. Since this means also that they are **local**, convergence order results on scaled grids $h\mathbb{Z}^n$ follow as well, albeit their proofs are made much more complicated by the lack of compact support of $\psi$ (only algebraic decay in comparison with the compact support of multivariate splines). This work, in turn, motivated [122] where such $\psi$ are used to compute fundamental Lagrange functions for interpolation on (finite) cardinal grids. A Gauss–Seidel type algorithm is used to invert the matrix $\{\psi(j-k)\}$, where $j$ and $k$ range over a finite subset of $\mathbb{Z}^n$. The matrix is amenable to this approach because the $\psi$ decay in fact quite fast (linked to the aforementioned polynomial recovery). Faster computation can be obtained by working out $\psi$ as a linear combination of translates of the multiquadric function explicitly. There is even a link in this work with an optimization package [119] that Mike wrote, because the absolute sum of the off-diagonal elements of the Gauss–Seidel matrix was minimized subject to a normalization condition and the coefficient constraints that give algebraic decay. The first step from equally spaced centres to unequally spaced ones was taken again by Mike Powell in [126], where a careful study of the univariate spaces generated by multiquadric translates led to a creation of quasi-interpolants by such radial functions with unequally spaced $x_j$ in one dimension. The spaces created by the radial function's translates and by the $\psi$s are quite different (both are completely described in [126]) and the latter is shown to contain all linear polynomials. Central to this work is the description of the $\psi$ functions that generate the quasi-interpolants by the Peano kernel theorem, the Peano kernel being the second derivative of the multiquadric function.

Focussing further on multiquadric interpolation, Mike studied in [130] its approximational accuracy on the unit interval when the centres are $h \cdot j$, $j = 0, \ldots, m$, where $c = h = 1/m$. Accuracy for Lipschitz continuous $f$ of $O(h)$ is established when $h \to 0$. His error estimate, which is valid uniformly on the whole interval, relies on carefully considering the size of the elements of the inverse of the interpolation matrix, including the boundary elements. If two conditions on $f'$ at 0 and 1 are satisfied additionally, then $O(h^2)$ accuracy can be shown.

The work of [126] has been carried further by Beatson and Powell [134] by studying three different quasi-interpolants with unequally spaced $x_j$. The first quasi-interpolant is in the $(m+1)$-dimensional space spanned by $\{\phi(\cdot - x_j)\}_{j=1}^m \cup \{\text{constant functions}\}$. The second is in the space spanned just by the $m$ translates of $\phi$. Quasi-interpolant number three is in the same space, enlarged by linear polynomials. The orders of accuracy obtained for the three quasi-interpolants are $(1+h^{-1}c)\omega_f(h)$, where $\omega_f$ is the modulus of continuity of $f$, and

$$h = \max_{2 \le j \le m}(x_j - x_{j-1}), \quad c\{|f(x_1) + f(x_m)|\}(x_m - x_1)^{-1} + (1+h^{-1}c)\omega_f(h)$$

and

$$\tfrac{1}{4}L\Big(c^2(1+2\log(1+(x_m - x_1)/c)) + \tfrac{1}{2}h^2\Big),$$

respectively. For the last estimate, $f'$ is assumed to satisfy a Lipschitz condition with the Lipschitz constant $L$.

Of course, this quasi-interpolant is the most interesting one which provides the best (i.e. second order except for the $\log c$ term) accuracy. Still, the other quasi-interpolants deserve attention too: the second one because it is an approximant from the space spanned only by translates of the multiquadric function. This can be considered a very natural space, because one usually interpolates from just that space. Finally, the first one deserves attention because it can be written as

$$s(x) = \text{const} + \sum_{j=1}^{m} \lambda_j \phi(x - x_j), \qquad x \in [x_1, x_m],$$

with an extra condition to take up the additional degree of freedom

$$\sum_{j=1}^{m} \lambda_j = 0. \tag{1.1}$$

This is the form that corresponds naturally to the variational formulation of radial basis function approximants such as the thin-plate spline approximants.

In [133], the results of [130] are extended by considering interpolation with centres and data points as previously, using translates of the multiquadric function plus a general linear polynomial. The multiquadric parameter $c$ is always a positive multiple of $h$. In the article, various ways to take up these extra degrees of freedom in such a way that superlinear convergence $o(h)$ is obtained are suggested (the authors conjecture that this is in fact $O(h^2)$, as supported by numerical evidence and the results of the previous paragraph). If the added linear polynomial is zero, then one cannot obtain more than $O(h)$ convergence for general twice differentiable functions unless the function satisfies boundary conditions. If a constant is added to the multiquadric approximant (and the extra degree of freedom is taken up by requiring that the coefficients of the multiquadric functions sum to zero), then superlinear convergence to twice continuously differentiable $f$ is obtained if and only if $f'(0) = f'(1) = 0$. If a linear polynomial is added and additionally to (1.1) $\sum_{j=1}^{m} \lambda_j x_j = 0$ is required, then superlinear convergence to $f \in C^2([0,1])$ is obtained if and only if $f'(0) = f'(1) = f(1) - f(0)$.

Apart from providing these necessary and sufficient conditions for superlinear convergence, Beatson and Powell suggest several new ways to take up the extra degrees of freedom in such a way that superlinear convergence is always obtained for twice continuously differentiable approximands; there is a proof that this is indeed the case for one of the methods put forward. They include interpolating $f'$ at 0 and 1, interpolating $f$ at $\frac{1}{2}h$ and $1 - \frac{1}{2}h$, and minimizing the sum of squares of interpolation coefficients. The latter is the choice for which superlinear convergence is proved.

In [132], Mike's opus magnum, he summarizes and explains many recent developments, including nonsingularity results for interpolation, polynomial reproduction and approximation order results for quasi-interpolation and Lagrange interpolation on cardinal grids for classes of radial basis functions, including all of the ones mentioned above and thin-plate splines $\phi(r) = r^2 \log r$, inverse (reciprocal) multiquadrics, and several others. The localisation of Lagrange functions for cardinal interpolation is considered in great detail and several improvements of known approximation order results are given. Much like his earlier review papers and his book, this work also does not just summarize his and other authors' work, but offers simplifications, more clarity in the exposition and improvements of results.

A further nice connection between approximation and optimization techniques can be found in [147] where approximants $s : \mathbb{R}^2 \to \mathbb{R}^2$ are considered that are componentwise thin-plate splines. The goal is to find a mapping between two regions in $\mathbb{R}^2$, where certain control points and control *curves* are mapped to prescribed positions. Mapping control points to points with the TPS method is not hard, but a curve must be discretized and it is not clear whether the discretization is the same in the image region even though the curve retains its shape. Because TPS yields the interpolant of minimal second derivatives in the least-squares sense, there is already one optimizing feature in that approach. In this article, Mike uses once more the universal algorithm [119] to determine the optimal positions of the discrete points on the curve in the image. The idea is to minimize again the semi-norm of the interpolant which consists of the sum of the square-integrals of its second partial derivatives but now with respect to the positions of the points of the discretized curve. Precisely, if $f_i$, $g_i$, $i = 1, 2, \ldots, m$, are the required image values, the semi-norm of the TPS interpolant turns out to be

$$8\pi (f_i)^T \, \Phi(f_i) + 8\pi (g_i)^T \, \Phi(g_i), \tag{1.2}$$

where $\Phi = \{\phi(\|x_j - x_k\|_2)\}_{j,k=1}^m$, $\phi(r) = r^2 \log r$, and $x_i$ are the points in the original domain in $\mathbb{R}^2$.

If we only want to map points into points, i.e.

$$\{x_i\}_{i=1}^m \to \{(f_i, g_i)\}_{i=1}^m ,$$

then (1.2) is the minimal value of the semi-norm that can be obtained. If, for simplicity, all of the $\{x_i\}_{i=1}^m$ originate from the discretization of the curve in the original domain, (1.2) can again be minimized with respect to the $(f_i, g_i)$, subject to those points lying on the given curve in the image domain. In particular, they should lie in the same order on the curve as the $\{x_i\}_{i=1}^m$ which gives linear inequality constraints to the optimization procedure. It is useful to write the points in a parametric form for this purpose.

In [143], the most general (with respect to the choice of the domain of convergence) results with regard to the convergence of thin-plate splines are obtained. There are several prior articles about convergence of thin-plate spline interpolants to scattered data on domains in $\mathbb{R}^2$, but the domains have always been required to have at least Lipschitz continuous boundaries. Mike succeeds in proving convergence within any bounded domain. The speed of convergence shown is within a factor of $\log h$ ($h$ being the largest minimum distance between interpolation points and any points in the domain), the same as the best of earlier results. On top of this, he gets the best multiplicative constants for the error estimates for interpolation on a line or within a square or a triangle, i.e. when we measure the error of thin-plate interpolation between two, three or four data points, where in the latter case they form a square. The $\log h$ term is due to the fact that the point $x$ where we measure the error need not be in the convex hull of the centres (though it does need to be in their $h$-neighbourhood, due to the definition of $h$).

At the time of writing this article, Mike's latest work considers the efficient solution of the thin-plate spline interpolation problem for a large volume of data. A closely related problem is the efficient evaluation of a given linear combination $s(x)$ of translates of $\|x\|^2 \log \|x\|$ many times, e.g. for the rendering on a computer screen. These two issues are related because the conditional positive definiteness of the interpolation matrix makes the CG (conjugate gradients) algorithm a suitable tool to solve the interpolation equations. And, of course, the CG algorithm needs many function evaluations of $s(x)$. One approach for evaluating $s(x)$ uses truncated Laurent expansions [136], [138] of the thin-plate splines and collecting several terms $\|x-x_j\|^2 \log \|x-x_j\|$ for $\|x_j\| >> \|x\|$ into one expression in order to minimize the number of evaluations of the logarithm, a computationally expensive task. A principal part of the work involved with this idea is deciding which of the $\|x-x_j\|^2 \log \|x-x_j\|$ are lumped together. When done efficiently, however, the cost of this, plus the approximation of the lumps by single truncated Laurent expansions, is $O(\log m)$ for $m$ centres plus $O(m \log m)$ set-up cost, small in comparison with at least $10m$ operations for direct evaluation.

Another approach for computing thin-plate spline interpolants efficiently by Mike Powell and collaborators uses local Lagrange functions, i.e. Lagrange functions $L_j$ centred at $x_j$, say, that satisfy the Lagrange conditions $L_j(x_k) = \delta_{jk}$ only for several $x_k$ near to $x_j$. The approximant is then constructed by a multigrid-type algorithm that exploits the observation that these local Lagrange functions are good approximations to the full Lagrange functions. This is in recognition of the fact that, at least if the data form an infinite regular grid, the full Lagrange functions decay exponentially, i.e. are very well localized. Therefore it is feasible to compute the interpolant by an iterative method which at each stage makes a correction to the residual by subtracting multiples of those local Lagrange functions. The iteration attempts to reduce the residual by subtracting $\sum_{i=1}^{m}(s(x_i) - f_i)\ L_i(x)$ from $s$, where $L_i$ are the local Lagrange functions and $s$ is the previous approximation to the thin-plate spline interpolant. It turns out that the iteration converges in many test cases, because the spectral radius of the iteration matrix associated with this procedure is less than 1. In a later paper [148], a slightly different approach is used where the *coefficient* of each $\|x - x_j\|^2 \log \|x - x_j\|$ is approximated in each step of the iteration by a multiple of the leading coefficient of a local Lagrange function $L_i(x)$ (the multiplier being the residual $f_i - s(x_i)$). Therefore the correcting term to a prior approximation is now

$$\sum_{i=1}^{m} \|x - x_i\|^2 \log \|x - x_i\| \sum_{j \in \mathcal{L}_i} \mu_{ij}[s(x_j) - f_j]\ ,$$

where $\mathcal{L}_i \subset \{1, 2, \ldots, m\}$ is the set of centres used for the local Lagrange function $L_i$ and $\mu_{ij}$ are its coefficients. This correction is performed iteratively until the required accuracy is obtained. The multigrid idea comes into play in this method as an inner iteration within each stage of the updating algorithm already described. Namely the above iterations are expected to remove the very high frequency components from the error. Therefore, there is now an inner iteration like a fine to coarse sweep of multigrid, where the set of centres is thinned out consecutively, and the updates as above are performed on the thinner sets, until just few centres are left which have not yet been considered. For those centres the correction then consists of solving the interpolation problem exactly. A remarkable observation is that the number of such iterations to obtain a prescribed accuracy seems to depend only weakly on the number of data.

# 2 The contributions of Mike Powell to optimization

I first came across Mike when I was a PhD student at the University of Leeds. I had been fortunate enough to come across a report of Bill Davidon on a variable metric method (strange words to me in those days). Having established that it was much better than anything currently available, I had dashed off a short paper which I was about to send to the *Computer Journal*, then the prime outlet in the UK for articles on numerical analysis. At the same time Mike was booked to give us a seminar on what was probably his first method for unconstrained minimization. This by the way was an ingenious way of obtaining quadratic termination in a gradient method, which however was already about to be superseded. A week before the seminar, Mike phoned and asked if he could change his title: he had come across a report of a much better method that he would like to talk about. Typically Mike had extracted the essentials of the method from the mass of detail in Davidon's flow sheets, and had also implemented it on the IBM machine at Harwell which was much faster than our modest Ferranti Pegasus at Leeds. When he heard of my interest in the method, he generously offered to pool his work with mine. We added some more things on conjugacy and such, and so was born the DFP paper [8]. Mike was also instrumental in promulgating the good news: he stood up at a meeting in London where speakers were elaborating on the difficulties of minimizing functions of 10 variables and told them that he had coded a method which had solved problems in 100 variables without difficulty. Of course this revolutionized the discipline and this type of method (but with the BFGS formula) is the method of choice to this day.

Since that time our paths have crossed many times, most notably when Mike recruited me to work at Harwell from 1969 to 1973. Although this led to only one other joint publication, I very much benefitted from discussing my work with him, and he was especially good at exposing weak arguments or suggesting useful theoretical and practical possibilities. (Also we made a formidable partnership at table football!) I remember an incident when I first arrived at Harwell and was telling him of some code in which I had used double precision accumulation of scalar products. His reply was something along the lines of "why bother with that, if an algorithm is numerically stable you should not need to use double precision in order to get adequate accuracy". I think that this is good advice in the sense that you learn much about an algorithm when you develop it in single precision, which hopefully can be used to good effect in improving numerical stability at any level of precision. *Apropos* of this, I hear that the recent Ariane rocket disaster was caused by a software error arising from the use of double precision floating point arithmetic, emphasizing how important it is to pay attention to errors arising from finite precision.

A lot of interest, particularly in the mid 1960s, centred on methods for unconstrained minimization without derivatives, and Mike published an important paper [9] in this area in 1964. I recently listened to a review paper crediting Mike with discovering in this paper how to obtain quadratic termination with line searches without evaluating derivatives. In fact this idea dates back to a report due to Smith in 1962, which Mike used as the basis of his paper. Mike's contribution was to extend the algorithm by including extra line searches which would be unnecessary in the quadratic case, but would enable the method to work more effectively for nonquadratic functions, whilst retaining termination for quadratics. An important feature was the manipulation of a set of "pseudo-conjugate directions" with a criterion to prevent these directions from becoming linearly dependent. In practice however it turned out to be very difficult to beat the DFP algorithm (and later the BFGS algorithm) with finite difference approximations to derivatives, much to Mike's (and other people's) disappointment. Recently, derivative-free methods have become fashionable again, with Mike in the forefront of this research [151]. He has again emphasized the importance of adequately modelling the problem functions, as against using heuristic methods like Nelder and Mead which are still well used despite their obvious disadvantages. His ideas also include a criterion to keep interpolation points from becoming coplanar, which harks back to his 1964 paper.

As methods developed, interest switched away from quadratic termination, towards proving global and local termination results. Here Mike's own contribution has been immense, and I shall say more about this in what follows. Moreover results obtained by subsequent researchers often use techniques of proof developed by him. Nonetheless Mike also has his feet on the ground and is aware of the importance of good performance on test problems and in real applications. I was struck by a phrase of his from an early paper that the performance of a method of his "can best be described as lively"!. Mike is also very good at constructing examples which illustrate the deficiencies in algorithms and so help rectify them. For example the Gauss–Newton method for nonlinear least squares (and hence Newton's method for equations) with an $\ell_2$ line search was a popular method, especially in the early days. A so-called convergence proof existed, and the method was often used with the confidence of a good outcome. The small print in the proof that the computed search direction $\mathbf{s}$ should be bounded was not usually given much thought, although it was known that convergence could be slow if the Jacobian matrix was rank deficient at the solution. Mike showed that the situation was much more serious by devising an example for which, in exact arithmetic, the algorithm converges to a nonstationary point. This led Mike to suggest his *dog-leg algorithm* [34] (here his enthusiasm for golf is seen) in which the search trajectory consists of a step along the steepest descent direction, followed by a step to the Newton point. The inclusion of the steepest descent component readily allows convergence to a stationary point to be proved, an idea that was much copied in subsequent research.

This leads me on to talk about *trust region methods* and it was interesting that there was some discussion at the birthday conference as to the historical facts. Of course the idea of modifying the Hessian by a multiple of a unit matrix, so as to induce a bias towards steepest descent, appears in the much referenced papers by Levenberg and by Marquardt. Marquardt gives the equivalence with minimizing the model on a ball, and attributes it to Morrison in 1960. Mike's contribution, which also first appears in [34] and then in [37], is to use the step restriction as a primary heuristic, rather than as a consequence of adding in a multiple of the unit matrix. Also Mike suggests the now well accepted test for increasing and decreasing the radius of the ball, based on the ratio of actual to predicted reduction in the objective function. Thus the framework which Mike proposed is what is now very much thought of as the prototypical trust region method. The term *trust region* was only coined some years later, possibly by John Dennis. In passing, the paper [37] was also notable for the introduction of what came to be known as the PSB (Powell–Symmetric–Broyden) updating formula, derived by an elegant iterative process. The variational properties of this formula were discovered a little later (by John Greenstadt) and subsequently led to the important work on the sparse PSB algorithm by Philippe Toint, who generously acknowledges Mike's contribution.

Talking about the value of small examples reminds me of another case where Mike removes the last shred of respectability from an algorithm. The idea of minimization by searching along the coordinate directions in turn dates back who knows when, and is known to exhibit slow convergence in practice. Mike's example [47] showed that it could also converge in exact arithmetic to a nonstationary point, and this removed the last reason for anyone being tempted to use the algorithm.

Without doubt, the conjecture that attracted the most interest in the 1960s was that of whether the DFP algorithm could converge to a non-stationary point. Mike accepted a bet of 1 shilling (£0.05) with Philip Wolfe that he (Mike) would solve the problem by some date. Although the method is a descent method, the result is anything but trivial, since the Hessian approximation may become singular or unbounded in the limit. Mike finally produced a convergence proof for strictly convex functions and exact line searches [38], that was a tour de force of analysis. I remember checking through it for him, a job which took me several days and left me thoroughly daunted at the complexity of what had been achieved. Mike later went on to prove [62] that the BFGS algorithm would converge for convex functions and an inexact line search, a more elegant result that has influenced most subsequent work on this topic. The conjecture is still open for nonconvex functions, so someone can still make a name for themselves here.

Another issue over many years has been to discuss the relative merits of different formulae in the Broyden family, particularly the DFP and BFGS formulae. Great excitement was caused at a Dundee meeting in 1971 when Lawrence Dixon introduced, in a rather peripheral way, his remarkable result for nonquadratic functions that all Broyden family updates generate identical points if exact line searches are used. Mike could hardly believe this and cross-examined Lawrence closely in question time. He then retired to his room to determine whether or not the result was indeed true. Having ascertained that it was, it is characteristic of Mike that he simplified Lawrence's proof and subsequently gave great credit to Lawrence for having made the discovery. One phenomenon of particular interest is the fact that for inexact line searches, the DFP formula can be much inferior to the BFGS formula, and various not very convincing explanations were advanced. One of my favourite Mike papers [104] is the one in which he analyses a very simple case of a two-variable quadratic and does a worst case analysis. The solution of the resulting recurrence relations is very neat, and comes out with a remarkably simple result about how the DFP formula behaves, which I think provides a very convincing explanation of the phenomenon.

Another fundamental algorithm is the conjugate gradient method, particularly in its form (Fletcher and Reeves) for nonquadratic optimization. Incidentally I would like to say here that Colin Reeves was my supervisor and not, as someone once suggested, my PhD student! It was a great loss to numerical analysis when he decided to become a computer scientist, and I had the greatest respect for his abilities. An important issue was the relative merits of the Fletcher–Reeves and Polak–Ribière formulae. Mike showed [67] that the PR formula was usually better in practice, but later [99] that the FR formula allowed a global convergence result, a result later extended by Al-Baali to allow inexact line searches. Mike also showed in [99] that the PR formula can fail to converge and derived a remarkable counterexample with $n = 3$ in which the sequence $\{\mathbf{x}^{(k)}\}$ is bounded and has six accumulation points, none of which is stationary. However I think Mike's most telling result

[64] is for quadratics, namely that if the method is started from an arbitrary descent direction, then either termination occurs, or the rate of convergence is linear, the latter possibility occurring with probability 1. This is bad news for the use of conjugate gradient methods in situations where the sequence must be broken, such as active set methods or limited memory methods, since it forces a restart in order to avoid linear convergence.

Turning to constrained optimization, Mike wrote an important paper [30] in 1969 that originated the idea of augmented Lagrangian penalty functions, at about the same time as a related paper by Hestenes on multiplier methods. Mike introduced the idea in a different way by making shifts $\theta_i$ to the constraint functions. After each minimization of the penalty function the shifts would be adjusted so as to get closer to the solution of the constrained problem. I seem to remember a talk in which Mike described how he was led to the idea when acting as a gunnery officer, presumably during a period of national service. Adjusting the $\theta_i$ parameters is analogous to the adjustment in elevation of a gun based on how close the previous shell lands from the target. One can have no doubts that Mike's gunnery crew would be the most successful in the whole battery! This method was a considerable improvement on the unshifted penalty function method, and led to considerable interest, and indeed is still used as a merit function in some nonlinear programming codes.

Mike also played an important part in the development of sequential quadratic programming (SQP) algorithms. Although the basic idea dates back to Wilson in 1963, and had been reviewed by Beale and others, the method had not attracted a great deal of interest. This was due to various factors, such as a reluctance to use expensive QP subproblems, the issue of how to update an approximation to the potentially indefinite Hessian of the Lagrangian, and what to do about infeasible QP subproblems. Mike [72] published an idea that handled the latter difficulty and got round the indefinite Hessian problem by ignoring it – or more accurately by suggesting a modification of the BFGS formula that could be used to update a positive definite approximation. However I think his greatest contribution was to popularize the method; which he did at conferences in 1977 at Dundee, Madison and Paris. His Madison paper [73] contained a justification of the idea of keeping a positive definite Hessian approximation. The term SQP was coined I think at the Cambridge NATO ASI in 1981 and the method has remained popular since that time, and is now frequently used in applications. Mike also contributed to a paper [87] on the *watchdog technique*, a title which I suspect is due to Mike, which popularized one way of avoiding difficulties such as the Maratos effect which might slow down the local convergence of SQP.

Since the early days at Harwell, Mike has diligently provided computer code so that the fullest use can be made of his ideas (although his style of programming is not what one might call structured!). It was natural therefore that he should write a QP code to support his SQP work, and he became a strong advocate [102] of the ideas in the Goldfarb–Idnani algorithm. These ideas also surfaced in his TOLMIN code for general linear constraint programming [121], and his ideas for updating conjugate directions in the BFGS formula [110]. The latter contains a particularly elegant idea for stacking up an ordered set of conjugate directions in the working matrix $Z$ as the algorithm proceeds. These ideas may not have caught on to the same extent as other papers that he has written, perhaps because they are not addressed towards large sparse calculations which are now attracting most attention. Nonetheless they contain a fund of good ideas that merit study. However, Mike has made other contributions to sparse matrix algebra, including the CPR algorithm [51] for estimating sparse Jacobian algorithms by finite differences in a very efficient way, and an extension of the same idea [75] to sparse Hessian updates.

More recently Mike has confirmed his ability to make advances in almost all areas of optimization. The Karmarkar interior point method for linear programming has spawned thousands of papers, but Mike nonetheless managed to make a unique contribution by constructing an example [125] showing that the algorithm can perform very badly as the number of constraints increases in a simple semi-infinite programming problem. Mike also continues to provoke the interior point community by his oft-expressed dislike for the need to use logarithms as a means of solving linear programming problems! Despite holding these views, Mike was nonetheless able to make an outstanding theoretical breakthrough in the convergence analysis of the shifted log barrier method for linear programming [145].

Much has been said about Mike as a formidable competitor in all that he he does, but I would like to close with a little reminiscence that sheds a different light. Mike has always been a faithful supporter of our biennial Dundee conferences, and has contributed a great deal to the success of these meetings. At one of these meetings, my Syrian PhD student Mehi Al-Baali was scheduled to give a talk on his work. As is the way of things for a local PhD student, he had been given a very unfavourable slot late on Friday afternoon. Mike however had to travel abroad the next day and so was planning to leave earlier in the afternoon to make the long eight-hour or so journey by road to Cambridge, to snatch a few hours sleep before his flight. When he saw Mehi's disappointment that he would not be present at the talk, he immediately agreed to postpone his departure and stay on a extra couple of hours or so for the talk. This is typical of the many kindnesses that Mike has shown to others, and I shall always remember him for that gesture. Happy birthday Mike!

# M.J.D. Powell's publications


[1] "Ground-state splitting for $d^5\,{}^6S$ ions in a cubic field" (with J.R. Gabriel and D.F. Johnston), *Phys. Rev. Letters* **5**, 145–146 (1960).

[2] "A scheme for handling functions on a computer", AERE Report No. R.3830, Harwell Laboratory (1961).

[3] "A calculation of the ground state splitting for $Mn^{++}$ ions in a cubic field" (with J.R. Gabriel and D.F. Johnston), *Proc. Roy. Soc. A* **264**, 503–515 (1961).

[4] "The crystalline field parameters for dysprosium ethyl sulphate" (with R. Orbach), *Proc. Phys. Soc.* **78**, 753–758 (1961).

[5] "An iterative method for finding stationary values of a function of several variables", *Computer Journal* **5**, 147–151 (1962).

[6] "Roothaan's procedure for solving the Hartree–Fock equation", in *Numerical Solution of Ordinary and Partial Differential Equations* (L. Fox, ed.), Pergamon Press (Oxford), 197–202 (1962).

[7] "Crystalline-field splittings in cerous magnesium nitrate" (with M.J.M. Leask, R. Orbach and W.P. Wolf), *Proc. Roy. Soc. A* **272**, 371–386 (1963).

[8] "A rapidly convergent descent method for minimization" (with R. Fletcher), *Computer Journal* **6**, 163–168 (1963).

[9] "An efficient method for finding the minimum of a function of several variables without calculating derivatives", *Computer Journal* **7**, 155–162 (1964).

[10] "The volume internal to three intersecting hard spheres", *Molecular Physics* **7**, 591–592 (1964).

[11] "A method for minimizing a sum of squares of nonlinear functions without calculating derivatives", *Computer Journal* **7**, 303–307 (1965).

[12] "Necessary conditions for a minimax approximation" (with A.R. Curtis), *Computer Journal* **8**, 358–361 (1966).

[13] "On the convergence of exchange algorithms for calculating minimax approximations" (with A.R. Curtis), *Computer Journal* **9**, 78–80 (1966).

[14–17] "The general theory of linear approximation" (with D.C. Handscomb and D.F. Mayers), "The exchange algorithm on a discrete point set", "Theory of general nonlinear minimax approximation" and "An introduction to epsilon entropy", in *Methods of Numerical Approximation* (D.C. Handscomb, ed.), Pergamon Press (Oxford), 61–71, 73–81, 155–162 and 183–190 (1966).

[18] "Minimization of functions of several variables", in *Numerical Analysis: An Introduction* (J. Walsh, ed.), Academic Press (London), 143–157 (1966).

[19] "Weighted uniform sampling—a Monte Carlo technique for reducing variance" (with J. Swann), *J. Inst. Maths. Applics.* **2**, 228–236 (1966).

[20] "On the maximum errors of polynomial approximations defined by interpolation and by least squares criteria", *Computer Journal* **9**, 404–407 (1967).

[21] "Error analysis of equal-interval interpolation by cubic splines" (with A.R. Curtis), AERE Report 5600, Harwell Laboratory (1967).

[22] "Using cubic splines to approximate functions of one variable to prescribed accuracy" (with A.R. Curtis), AERE Report No. R.5602, Harwell Laboratory (1967).

[23] "On best $L_2$ spline approximations", in *Numerische Mathematik, Differentialgleichungen Approximationstheorie* Birkhäuser-Verlag (Basel), 317–339 (1968).

[24] "On the calculation of orthogonal vectors", *Computer Journal* **11**, 302–304 (1968).

[25] "A Fortran subroutine to invert a rectangular matrix of full rank", AERE Report No. R.6072, Harwell Laboratory (1969).

[26] "A theorem on rank one modifications to a matrix and its inverse", *Computer Journal* **12**, 288–290 (1969).

[27] "A comparison of spline approximations with classical interpolation methods", in *Proceedings of the IFIP 1968 Congress,* North-Holland (Amsterdam), 95–98 (1969).

[28] "On applying Householder transformations to linear least squares problems" (with J.K. Reid), in *Proceedings of the IFIP 1968 Congress,* North-Holland (Amsterdam), 122–126 (1969).

[29] "The local dependence of least squares cubic splines", *SIAM J. Numer. Anal.* **6**, 398–413 (1969).

[30] "A method for nonlinear constraints in minimization problems", in *Optimization* (R. Fletcher, ed.), Academic Press (London), 283–297 (1969).

[31] "A survey of numerical methods for unconstrained optimization", *SIAM Review* **12**, 79–97 (1970).

[32] "Curve fitting by splines in one variable", in *Numerical Approximation to Functions and Data* (J.G. Hayes, ed.), Athlone Press (London), 65–83 (1970).

[33] "Rank one methods for unconstrained optimization", in *Integer and Nonlinear Programming* (J. Abadie, ed.), North-Holland (Amsterdam), 139–156 (1970).

[34] "A hybrid method for nonlinear equations", in *Numerical Methods for Nonlinear Equations* (P. Rabinowitz, ed.), Gordon and Breach (London), 87–114 (1970).

[35] "A Fortran subroutine for solving systems of nonlinear algebraic equations", in *Numerical Methods for Nonlinear Equations* (P. Rabinowitz, ed.), Gordon and Breach (London), 115–161 (1970).

[36] "A Fortran subroutine for unconstrained minimization requiring first derivatives of the objective function", AERE Report No. R.6469, Harwell Laboratory (1970).

[37] "A new algorithm for unconstrained optimization", in *Nonlinear Programming* (J.B. Rosen, O.L. Mangasarian and K. Ritter, eds), Academic Press (New York), 31–65 (1970).

[38] "On the convergence of the variable metric algorithm", *J. Inst. Maths. Applics.* **7**, 21–36 (1971).

[39] "Recent advances in unconstrained optimization", *Math. Programming* **1**, 26–57 (1971).

[40] "Some properties of the variable metric algorithm", in *Numerical Methods for Nonlinear Optimization* (F.A. Lootsma, ed.), Academic Press (London), 1–17 (1972).

[41] "The differential correction algorithm for rational $L_\infty$ approximation" (with I. Barrodale and F.D.K. Roberts), *SIAM J. Numer. Anal.* **9**, 493–504 (1972).

[42] "Quadratic termination properties of minimization algorithms: 1. Statement and discussion of results", *J. Inst. Maths. Applics.* **10**, 333–342 (1972).

[43] "Quadratic termination properties of minimization algorithms: 2. Proofs of theorems", *J. Inst. Maths. Applics.* **10,** 343–357 (1972).

[44] "Problems related to unconstrained optimization", in *Numerical Methods for Unconstrained Optimization* (W. Murray, ed.), Academic Press (London), 29–55 (1972).

[45] "A Fortran subroutine for drawing a curve through a sequence of data points" (with S. Marlow), AERE Report No. R.7092, Harwell Laboratory (1972).

[46] "A Fortran subroutine for calculating a cubic spline approximation to a given function", AERE Report No. R.7308, Harwell Laboratory (1972).

[47] "On search directions for minimization algorithms", *Math. Programming* **4**, 193–201 (1973).

[48] "A Fortran subroutine for plotting a cubic spline function" (with S. Marlow), AERE Report No. R.7470, Harwell Laboratory (1973).

[49] "Numerical methods for constrained optimization", *IMA Bulletin* **9**, 326–327 (1973).

[50] "Unconstrained minimization algorithms without computation of derivatives", *Boll. della Unione Matematica Italiana* **9**, 60–69 (1974).

[51] "On the estimation of sparse Jacobian matrices" (with A.R. Curtis and J.K. Reid), *J. Inst. Maths. Applics.* **13**, 117–119 (1974).

[52] "Unconstrained minimization and extensions for constraints", in *Mathematical Programming in Theory and Practice* (P.L. Hammer and G. Zoutendijk, eds), North-Holland (Amsterdam), 31–79 (1974).

[53] "Piecewise quadratic surface fitting for contour plotting", in *Software for Numerical Mathematics* (D.J. Evans, ed.), Academic Press (London), 253–271 (1974).

[54] "On the modification of $LDL^T$ factorizations" (with R. Fletcher), *Maths. of Comp.* **28**, 1067–1087 (1974).

[55] "Introduction to constrained optimization", in *Numerical Methods for Constrained Optimization* (P.E. Gill and W. Murray, eds), Academic Press (London), 1–28 (1974).

[56] "On the convergence of cyclic Jacobi methods" (with K.W. Brodlie), *J. Inst. Maths. Applics.* **15**, 279–287 (1975).

[57] "Convergence properties of a class of minimization algorithms", in *Nonlinear Programming 2* (O.L. Mangasarian, R.R. Meyer and S.M. Robinson, eds), Academic Press (New York), 1–27 (1975).

[58] "A view of minimization algorithms that do not require derivatives", *ACM Trans. Math. Software* **1**, 97–107 (1975).

[59] "The minimax solution of linear equations subject to bounds on the variables", in *Proc. of the Fourth Manitoba Conference on Numerical Mathematics,* Utilitas Matematica Publishing Inc. (Winnipeg), 53–107 (1975).

[60] "A Fortran subroutine that calculates the minimax solution of linear equations subject to bounds on the variables" (with K. Madsen), AERE Report No. R.7954, Harwell Laboratory (1975).

[61] "A view of unconstrained optimization", in *Optimization in Action* (L.C.W. Dixon, ed.), Academic Press (London), 117–152 (1976).

[62] "Some global convergence properties of a variable metric algorithm for minimization without exact line searches", in *Nonlinear Programming SIAM-AMS Proceedings, Vol. IX* (R.W. Cottle and C.E. Lemke, eds), American Mathematical Society (Providence), 53–72 (1976).

[63] "Optimal interpolation" (with P.W. Gaffney), in *Numerical Analysis Dundee 1975, Lecture Notes in Mathematics No. 506* (G.A. Watson, ed.), Springer-Verlag (Berlin), 90–99 (1976).

[64] "Some convergence properties of the conjugate gradient method", *Math. Programming* **11**, 42–49 (1976).

[65] "A Fortran subroutine for plotting the part of a conic that is inside a given triangle" (with S. Marlow), AERE Report No. R.8336, Harwell Laboratory (1976).

[66] "Quadratic termination properties of Davidon's new variable metric algorithm", *Math. Programming* **11**, 141–147 (1977).

[67] "Restart procedures for the conjugate gradient method", *Math. Programming* **12**, 241–254 (1977).

[68] "Numerical methods for fitting functions of two variables", in *The State of the Art in Numerical Analysis* (D.A.H. Jacobs, ed.), Academic Press (London), 563–604 (1977).

[69] "Piecewise quadratic approximations on triangles" (with M.A. Sabin), *ACM Trans. Math. Software* **3**, 316–325 (1977).

[70] "A technique that gains speed and accuracy in the minimax solution of overdetermined linear equations" (with M.J. Hopper), in *Mathematical Software 3* (J.R. Rice, ed.), Academic Press (New York), 15–33 (1977).

[71] "Algorithms for nonlinear constraints that use Lagrangian functions", *Math. Programming* **14**, 224–248 (1978).

[72] "A fast algorithm for nonlinearly constrained optimization calculations", in *Numerical Analysis Dundee 1977, Lecture Notes in Mathematics No. 630* (G.A. Watson, ed.), Springer-Verlag (Berlin), 144–157 (1978).

[73] "The convergence of variable metric methods for nonlinearly constrained optimization calculations", in *Nonlinear Programming 3* (O.L. Mangasarian, R.R. Meyer and S.M. Robinson, eds), Academic Press (New York), 27–63 (1978).

[74] "Variable metric methods for constrained optimization", in *Computing Methods in Applied Sciences and Engineering 1977 (1), Lecture Notes in Mathematics No. 704* (R. Glowinski and J.L. Lions, eds), Springer-Verlag (Berlin), 62–72 (1979).

[75] "On the estimation of sparse Hessian matrices" (with Ph.L. Toint), *SIAM J. Numer. Anal.* **16**, 1060–1074 (1979).

[76–77] "Gradient conditions and Lagrange multipliers in nonlinear programming" and "Variable metric methods for constrained optimization", in *Nonlinear Optimization Theory and Algorithms* (L.C.W. Dixon, ed.), E. Spedicato and G. Szegő, Birkhäuser (Boston), 201–220 and 279–294 (1980).

[78] "Optimization algorithms in 1979", in *Optimization Techniques, Lecture Notes in Control and Information Sciences No. 22* (K. Iracki, K. Malanowski and S. Walukiewicz, eds), Springer-Verlag (Berlin), 83–98 (1980).

[79] "A discrete characterization theorem for the discrete $L_1$ linear approximation problem" (with F.D.K. Roberts), *J. Approx. Theory* **30**, 173–179 (1980).

[80] "A note on quasi-Newton formulae for sparse second derivative matrices", *Math. Programming* **20**, 144–151 (1981).

[81] *Approximation Theory and Methods,* Cambridge University Press (Cambridge) (1981).

[82] "An example of cycling in a feasible point algorithm", *Math. Programming* **20**, 353–357 (1981).

[83] "On the decomposition of conditionally positive-semidefinite matrices" (with D.H. Martin and D.H. Jacobson), *Linear Algebra Applics* **39**, 51–59 (1981).

[84] "On the A-acceptability of rational approximations to the exponential function" (with A. Iserles), *IMA J. Numer. Anal.* **1**, 241–251 (1981).

[85] "An upper triangular matrix method for quadratic programming", in *Nonlinear Programming 4* (O.L. Mangasarian, R.R. Meyer and S.M. Robinson, eds), Academic Press (New York), 1–24 (1981).

[86] "The Shanno–Toint procedure for updating sparse symmetric matrices" (with Ph.L. Toint), *IMA J. Numer. Anal.* **1**, 403–413 (1981).

[87] "The watchdog technique for forcing convergence in algorithms for constrained optimization" (with R.M. Chamberlain, C. Lemaréchal and H.C. Pedersen), *Math. Programming Studies* **16**, 1–17 (1982).

[88] *Nonlinear Optimization 1981* (editor), Academic Press (London) (1982).

[89] "Data smoothing by divided differences" (with M.P. Cullinan), in *Numerical Analysis Proceedings, Dundee 1981, Lecture Notes in Mathematics No. 912* (G.A. Watson, ed.), Springer-Verlag (Berlin), 26–37 (1982).

[90] "Extensions to subroutine VFO2AD", in *System Modeling and Optimization, Lecture Notes in Control and Information Sciences No. 38* (R.F. Drenick and F. Kozin, eds), Springer-Verlag (Berlin), 529–538 (1982).

[91] "Algorithms for constrained and unconstrained optimization calculations", in *Current Developments in the Interface: Economics, Econometrics, Mathematics* (M. Hazewinkel and H.G. Rinnooy Kan, eds), Reidel (Dordrecht), 293–310 (1982).

[92] "VMCWD: A Fortran subroutine for constrained optimization", Report No. DAMTP 1982/NA4, University of Cambridge (1982).

[93] "The convergence of variable metric matrices in unconstrained optimization" (with R-P. Ge), *Math. Programming* **27**, 123–143 (1983).

[94] "Variable metric methods for constrained optimization", in *Mathematical Programming: The State of the Art, Bonn 1982* (A. Bachem, M. Grötschel and B. Korte, eds), Springer-Verlag (Berlin), 288–311 (1983).

[95] "General algorithms for discrete nonlinear approximation calculations", in *Approximation Theory IV* (C.K. Chui, L.L. Schumaker and J.D. Ward, eds), Academic Press (New York), 187–218 (1983).

[96] "ZQPCVX: A Fortran subroutine for convex quadratic programming", Report No. DAMTP 1983/NA17, University of Cambridge (1983).

[97] "On the global convergence of trust region algorithms for unconstrained minimization", *Math. Programming* **29**, 297–303 (1984).

[98] "Conditions for superlinear convergence in $\ell_1$ and $\ell_\infty$ solutions of overdetermined nonlinear equations" (with Y. Yuan), *IMA J. Numer. Anal.* **4**, 241–251 (1984).

[99] "Nonconvex minimization calculations and the conjugate gradient method", in *Numerical Analysis Proceedings, Dundee 1983, Lecture Notes in Mathematics No. 1066* (G.A. Watson, ed.), Springer-Verlag (Berlin), 122–141 (1984).

[100] "On the rate of convergence of variable metric algorithms for unconstrained optimization", in *Proceedings of the International Congress of Mathematicians Warsaw 1983* (Z. Ciesielski and C. Olech, eds), North-Holland (Amsterdam), 1525–1539 (1984).

[101] "An application of Gaussian elimination to interpolation by generalised rational functions" (with T. Håvie), in *Rational Approximation and Interpolation, Lecture Notes in Mathematics No. 1105* (P.R. Graves-Morris, E.B. Saff and R.S. Varga, eds), Springer-Verlag (Berlin), 442–452 (1984).

[102] "On the quadratic programming algorithm of Goldfarb and Idnani", *Math. Programming Studies* **25**, 46–61 (1985).

[103] "The performance of two subroutines for constrained optimization on some difficult test problems," in *Numerical Optimization 1984* (P.T. Boggs, R.H. Byrd and R.B. Schnabel, eds), SIAM Publications (Philadelphia), 160–177 (1985).

[104] "How bad are the BFGS and DFP methods when the objective function is quadratic?", *Math. Programming* **34**, 34–47 (1986).

[105] "Convergence properties of algorithms for nonlinear optimization", *SIAM Review* **28**, 487–500 (1986).

[106] "A recursive quadratic programming algorithm that uses differentiable exact penalty functions" (with Y. Yuan), *Math. Programming* **35**, 265–278 (1986).

[107] "Radial basis functions for multivariable interpolation: a review", in *Algorithms for Approximation* (J.C. Mason and M.G. Cox, eds), Oxford University Press (Oxford), 143–167 (1987).

[108] "The differential correction algorithm for generalized rational functions" (with E.W. Cheney), *Constructive Approximation* **3**, 249–256 (1987).

[109] "On error growth in the Bartels–Golub and Fletcher–Matthews algorithms for updating matrix factorizations," *Linear Algebra Applics* **88–89**, 597–621 (1987).

[110] "Updating conjugate directions by the BFGS formula", *Math. Programming* **38**, 29–46 (1987).

[111] *The State of the Art in Numerical Analysis* (editor with A. Iserles), Oxford University Press (Oxford) (1987).

[112] "Methods for nonlinear constraints in optimization calculations", in *The State of the Art in Numerical Analysis* (A. Iserles and M.J.D. Powell, eds), Oxford University Press (Oxford), 325–357 (1987).

[113] "A biographical memoir of Evelyn Martin Lansdowne Beale", *Biographical Memoirs of Fellows of the Royal Society* **33**, 23–45 (1987).

[114] "Radial basis function approximation to polynomials", in *Numerical Analysis 1987* (D.F. Griffiths and G.A. Watson, eds), Longman Scientific and Technical (Burnt Mill), 223–241 (1988).

[115] *Mathematical Models and their Solutions: Contributions to the Martin Beale Memorial Symposium* (editor), *Math. Programming B* **42(1)** (1988).

[116] "An algorithm for maximizing entropy subject to simple bounds", *Math. Programming B* **42**, 171–180 (1988).

[117] "A review of algorithms for nonlinear equations and unconstrained optimization", in *ICIAM '87: Proceedings of the First International Conference on Industrial and Applied Mathematics* (J. McKenna and R. Temam, eds), SIAM Publications (Philadelphia), 220–232 (1988).

[118] "QR factorization for linear least squares problems on the hypercube" (with R.M. Chamberlain), *IMA J. Numer. Anal.* **8**, 401–413 (1988).

[119] "TOLMIN: A Fortran package for linearly constrained optimization calculations", Report No. DAMTP 1989/NA2, University of Cambridge (1989).

[120] "On a matrix factorization for linearly constrained optimization problems", in *Applications of Matrix Theory* (M.J.C. Gover and S. Barnett, eds), Oxford University Press (Oxford), 83–100 (1989).

[121] "A tolerant algorithm for linearly constrained optimization calculations", *Math. Programming B* **45**, 547–566 (1989).

[122] "Radial basis function interpolation on an infinite regular grid" (with M.D. Buhmann), in *Algorithms for Approximation II* (J.C. Mason and M.G. Cox, eds), Chapman and Hall (London), 146–169 (1990).

[123] "Algorithms for linearly constrained optimization calculations", in *Proceedings of the Third IMSL User Group Europe Conference* (M. Vaccari, ed.), IMSL (Houston), paper A-1, 1–17 (1990).

[124] "The updating of matrices of conjugate directions in optimization algorithms", in *Numerical Analysis 1989* (D.F. Griffiths and G.A. Watson, eds), Longman Scientific and Technical (Burnt Mill), 193–205 (1990).

[125] "Karmarkar's algorithm: a view from nonlinear programming", *IMA Bulletin* **26**, 165–181 (1990).

[126] "Univariate multiquadric approximation: reproduction of linear polynomials", in *Multivariate Approximation and Interpolation* (W. Haussmann and K. Jetter, eds), Birkhäuser Verlag (Basel), 227–240 (1990).

[127] "A trust region algorithm for equality constrained optimization " (with Y. Yuan), *Math. Programming A* **49**, 189–211 (1991).

[128] "Least squares smoothing of univariate data to achieve piecewise monotonicity" (with I.C. Demetriou), *IMA J. Numer. Anal.* **11**, 411–432 (1991).

[129] "The minimum sum of squares change to univariate data that gives convexity" (with I.C. Demetriou), *IMA J. Numer. Anal.* **11**, 433–448 (1991).

[130] "Univariate multiquadric interpolation: some recent results", in *Curves and Surfaces* (P.J. Laurent, A. Le Méhauté and L.L. Schumaker, eds), Academic Press (New York), 371–382 (1991).

[131] "A view of nonlinear optimization", in *History of Mathematical Programming: A Collection of Personal Reminiscences* (J.K. Lenstra, A.H.G. Rinnooy Kan and A. Schrijver, eds), North-Holland (Amsterdam), 119–125 (1991).

[132] "The theory of radial basis function approximation in 1990", in *Advances in Numerical Analysis II: Wavelets, Subdivision Algorithms and Radial Basis Functions* (W. Light, ed.), Oxford University Press (Oxford), 105–210 (1992).

[133] "Univariate interpolation on a regular finite grid by a multiquadric plus a linear polynomial" (with R.K. Beatson), *IMA J. Numer. Anal.* **12**, 107–133 (1992).

[134] "Univariate multiquadric approximation: quasi-interpolation to scattered data" (with R.K. Beatson), *Constructive Approximation* **8**, 275–288 (1992).

[135] "The complexity of Karmarkar's algorithm for linear programming", in *Numerical Analysis, 1991* (D.F. Griffiths and G.A. Watson, eds), Longman Scientific and Technical (Burnt Mill), 142–163 (1992).

[136] "Tabulation of thin plate splines on a very fine two-dimensional grid", in *Numerical Methods in Approximation Theory* **9** (D. Braess and L. Schumaker, eds), Birkhäuser Verlag (Basel), 221–244 (1992).

[137] "On the number of iterations of Karmarkar's algorithm for linear programming", *Math. Programming* **62**, 153–197 (1993).

[138] "Truncated Laurent expansions for the fast evaluation of thin plate splines", *Numerical Algorithms* **5**, 99–120 (1993).

[139] "Log barrier methods for semi-infinite programming calculations", in *Advances on Computer Mathematics and its Applications* (E.A. Lipitakis, ed.), World Scientific (Singapore), 1–21 (1993).

[140] "An iterative method for thin plate spline interpolation that employs approximations to Lagrange functions" (with R.K. Beatson), in *Numerical Analysis, 1993* (D.F. Griffiths and G.A. Watson, eds), Longman Scientific and Technical (Burnt Mill), 17–39 (1994).

[141] "A direct search optimization method that models the objective and constraint functions by linear interpolation", in *Advances in Optimization and Numerical Analysis* (S. Gomez and J-P. Hennart, eds), Kluwer Academic (Dordrecht), 51–67 (1994).

[142] "Some algorithms for thin plate spline interpolation to functions of two variables", in *Advances in Computational Mathematics: New Delhi, India* (H.P. Dikshit and C.A. Micchelli, eds), World Scientific (Singapore), 303–319 (1994).

[143] "The uniform convergence of thin plate spline interpolation in two dimensions", *Numerische Mathematik* **68**, 107–128 (1994).

[144] "A 'taut string algorithm' for straightening a piecewise linear path in two dimensions", Report No. DAMTP 1994/NA7, University of Cambridge, to appear in *IMA J. Num. Anal.*

[145] "Some convergence properties of the modified log barrier method for linear programming", *SIAM J. Optimization* **5**, 695–739 (1995).

[146] "An algorithm that straightens and smooths piecewise linear curves in two dimensions", in *Mathematical Methods for Curves and Surfaces* (M. Daehlen, T. Lyche and L.L. Schumaker, eds), Vanderbilt University Press (Nashville), 439–453 (1995).

[147] "A thin plate spline method for mapping curves into curves in two dimensions", in *Computational Techniques and Applications: CTAC95* (R.L. May and A.K. Easton, eds), World Scientific (Singapore), 43–57 (1996).

[148] "On multigrid techniques for thin plate spline interpolation in two dimensions" (with R.K. Beatson and G. Goodsell), in *The Mathematics of Numerical Analysis* (J. Renegar, M. Shub and S. Smale, eds), American Mathematical Society (Providence), 77–97 (1996).

[149] "A review of algorithms for thin plate spline interpolation in two dimensions", in *Advanced Topics in Multivariate Approximation* (F. Fontanella, K. Jetter and P.J. Laurent, eds), World Scientific (Singapore), 303–322 (1996).

[150] "A review of methods for multivariable interpolation at scattered data points", Report No. DAMTP 1996/NA11, University of Cambridge.

[151] "Least squares fitting to univariate data subject to restrictions on the signs of the second differences" (with I.C. Demetriou), this volume.

[152] "A new iterative algorithm for thin plate spline interpolation in two dimensions", *Annals of Numerical Mathematics* **4**, 519–527 (1997).

# MJDP–BCS Industrial Liaison: Applications to Defence Science

*Ian Barrodale*
*Cedric Zala*
Barrodale Computing Services Ltd, P.O. Box 3075, Victoria, B.C., Canada V8W 3W2

**Abstract**

Professor M.J.D. Powell has collaborated with Barrodale Computing Services Ltd (BCS) in our contract research and development work over the past several years. The topics have ranged from image processing and time series analysis to approximation theory and numerical analysis. The applications have been primarily in defence science (including underwater acoustics, mine countermeasures and nondestructive evaluation of materials), but also in areas involving natural resources (forest inventory mapping and seismic exploration). A selection of Prof. Powell's contributions to BCS in these fields is reviewed.

## 1 Introduction

The first-named author (IB) and Mike Powell have interacted professionally and socially (both on and off the golf course) for the past thirty years, having first met when Mike (then at Harwell) gave a talk at Liverpool University in 1966. Subsequently, Mike visited the University of Victoria, where he provided generous encouragement to IB and colleagues in their research on approximation algorithms. Since moving to Cambridge in 1976, Mike has continued to visit Victoria (often with his family), and after BCS was incorporated in 1978, this professional interaction focussed primarily on practical solutions to algorithmic problems motivated by requirements of BCS industrial software projects. Both authors have posed scientific computing problems for our academic consultant (Mike Powell), who has invariably contributed ideas, algorithms and software for their solution. This paper discusses, in the context of defence science, three such areas (high-resolution deconvolution, inversion, and image warping) where Mike Powell has contributed significantly to our technical successes at BCS.

# 2 Time series analysis

## 2.1 High-resolution deconvolution

In geophysical applications, a seismic trace $t$ of length $n$ is usually modelled as a convolution of a source wavelet $w$ of length $k$ with a spike series $s$ of length $m$, together with additive noise $r$:

$$t = w * s + r \qquad\qquad (n = m + k - 1).$$

Generalizing this model to other applications, and adopting matrix notation, the deconvolution problem is to determine a vector $\mathbf{s} = [s_1, s_2, \ldots, s_m]^T$ which minimizes $||\mathbf{Ws} - \mathbf{t}||$, where $\mathbf{t} = [t_1, t_2, \ldots, t_n]^T$, and the $n \times m$ convolutional matrix $\mathbf{W}$ is defined as

$$\mathbf{W} = \begin{bmatrix} w_1 & & & & \\ w_2 & w_1 & & 0 & \\ w_3 & w_2 & w_1 & & \\ w_4 & w_3 & w_2 & \ddots & \\ \vdots & w_4 & w_3 & \ddots & w_1 \\ w_k & \vdots & w_4 & \ddots & w_2 \\ & w_k & \vdots & \ddots & w_3 \\ & & w_k & \ddots & w_4 \\ & 0 & & \ddots & \vdots \\ & & & & w_k \end{bmatrix}$$

High-resolution deconvolution refers to cases where a sparse vector $\mathbf{s}$ (i.e., with isolated spikes) is sought; say, when analyzing physical data corresponding to depth estimates of geological layers, material interfaces, or cracks within an aircraft wing. Figure 2.1 shows an example of a sparse spike series (top), along with three types of wavelets. When this spike series is convolved with each of these wavelets, a low level of noise is added, and the resulting traces are deconvolved by the high-resolution $L_2$ deconvolution algorithm described below, the estimated spike series indicated in Figure 2.1 result.

In the development of suitable algorithms for high-resolution deconvolution, an important consideration is that this system of equations is frequently ill-conditioned in practice, particularly when the wavelet is band-limited[1] (as in case (c) of Figure 2.1). In this context, Milinazzo *et al.* (1987) showed that the condition number of $\mathbf{W}$ is proportional to $m^p$, where $p$ is the largest order of any zero in the Fourier spectrum of $w$.

Claerbout and Muir (1973) first drew attention to the possible use of the $L_1$ norm for calculating a sparse spike series, but our introduction to this

[1] i.e., its Fourier spectrum has one or more zeros.

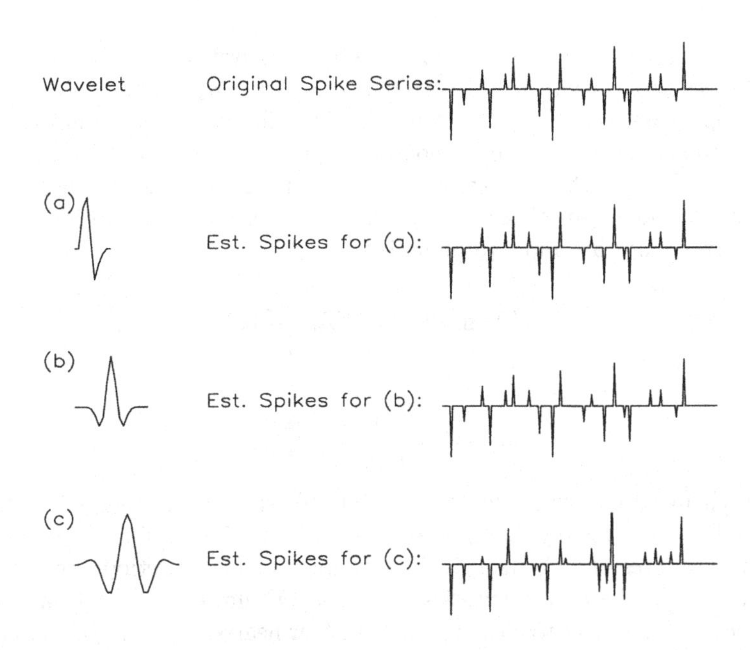


**Figure 2.1** Sparse deconvolution of traces formed using various types of wavelets: (a) asymmetric broadband wavelet; (b) symmetric broadband wavelet; (c) band-limited wavelet.

problem was provided by Taylor *et al.* (1979), who extracted spike series by minimizing $||\mathbf{Ws} - \mathbf{t}||_1 + \lambda||\mathbf{s}||_1$. As $\lambda$ increases, the use of the $L_1$ norm forces many spike amplitudes to be zero, thus yielding a sparse result. In practice, difficulties with this approach arise in the specification of an appropriate value for $\lambda$, and the consequent computational effort required in calculating a new solution for each trial value of $\lambda$ (typically $n > 250$, and $25 < k \ll m < n$). In any event, the inclusion here of the spike series term $||\mathbf{s}||$ prevents the true minimization of the residual error.

In the early 1980s, BCS developed a new approach to high-resolution deconvolution (first described in Chapman and Barrodale (1983)) by adapting the simplex method of linear programming to construct $\mathbf{s}$ in a manner which causes $||\mathbf{Ws} - \mathbf{t}||_1$ to decrease by the maximum amount possible as each new spike is inserted. Thus, in this one-at-a-time spike extraction algorithm, $\mathbf{s}^{(j+1)}$ is determined in the $(j+1)$th simplex iteration by minimizing $||\mathbf{Ws} - \mathbf{t}||_1$ subject to $\mathbf{s}$ having $(j+1)$ nonzero spikes, of which $j$ have the same positions as $\mathbf{s}^{(j)}$. Clearly, $||\mathbf{Ws}^{(j+1)} - \mathbf{t}||_1 \leq ||\mathbf{Ws}^{(j)} - \mathbf{t}||_1$, and iterations can be continued until this monotone decreasing behaviour slows down, or until a predetermined number of spikes has been extracted. A version of this algorithm was sold in 1981 to a major Canadian oil company, who incorporated it as a component in their seismic data analysis system.

In 1988, we asked Powell to examine this earlier work with a view towards developing an improved algorithm. His first observation was that the difficult problem of minimizing a nonconvex function with many local minima is unavoidable in these types of deconvolution calculations. However, Powell (1988) developed an algorithm implementing sparse deconvolution with the $L_2$ norm, which attempts to minimize

$$||\mathbf{Ws} - \mathbf{t}||_2^2 + \mathrm{TOL} \cdot \sum_{i=1}^{m} |s_i|^0,$$

where

$$|s_i|^0 = \begin{cases} 1, & s_i \neq 0 \\ 0, & s_i = 0 \end{cases} .$$

In each iteration, spikes are added, deleted, shifted and merged, until the error is small. Ideally, TOL is assigned a value that is similar to the final sum of squares of residuals. Various strategies and restrictions were devised by Powell in the implementation of this algorithm, e.g., to precompute the autocorrelation, and to consider only the four nearest spike neighbours in the calculations. For appropriate choice of TOL, this approach was found to produce sparse estimates for **s** which were often similar to those produced by the $L_1$ one-at-a-time procedure, but more than ten times faster. Again, examples of the spike series estimates produced by this deconvolution algorithm are shown in Figure 2.1.

In addition to its usefulness in geophysics, high-resolution deconvolution is also of benefit in the analysis of ultrasonic data for nondestructive evaluation (NDE) of materials. In the inspection of aircraft components, pulse-echo techniques are used, in which a transducer transmits a focussed pressure wave into the material and then measures the reflections. These data are often collected along parallel sets of scan lines, so that the data set is 3-D. From these data, images termed B-scans (a set of traces along one scan dimension) and C-scans (reflection amplitudes within a certain time gate $\Delta t$) can be produced, as indicated in Figure 2.2.

A specific example of the effectiveness of high-resolution deconvolution is its application to detection of defects in an aircraft steplap joint. This component is composed of a multiply-stepped titanium layer embedded within laminations of graphite-epoxy. Figure 2.3 shows a specially manufactured specimen with this structure, in which two defects were simulated by inserting rectangular Teflon inserts during fabrication.

A B-scan consisting of 256 traces, each containing 256 elements, was obtained for this specimen using a 5 MHz transducer. This B-scan is shown in the top left image of Figure 2.4. Because of the relatively long time extent of the pressure wave generated by the transducer, the data are poorly resolved in the vertical dimension. In general the spreading of the reflections caused

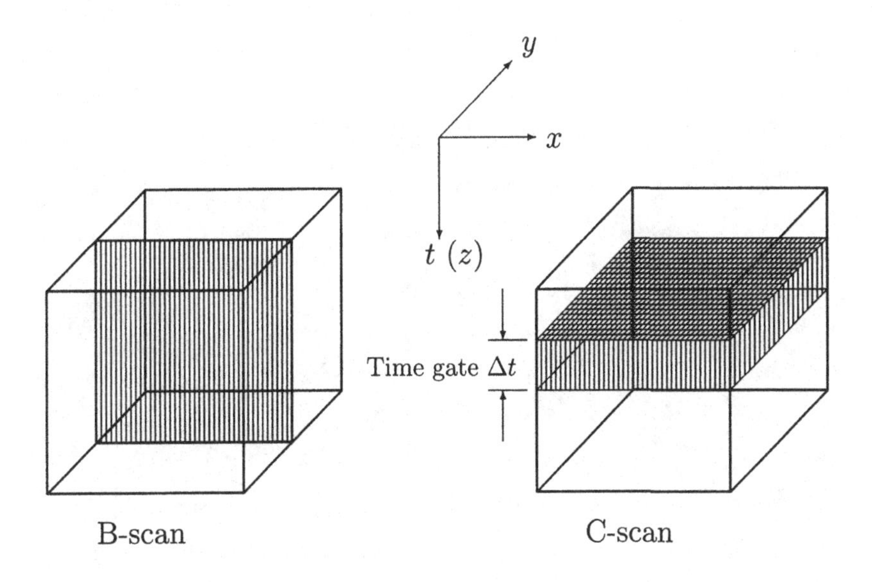


**Figure 2.2** Illustration of the formation of B-scans and C-scans from 3-D ultrasonic data.

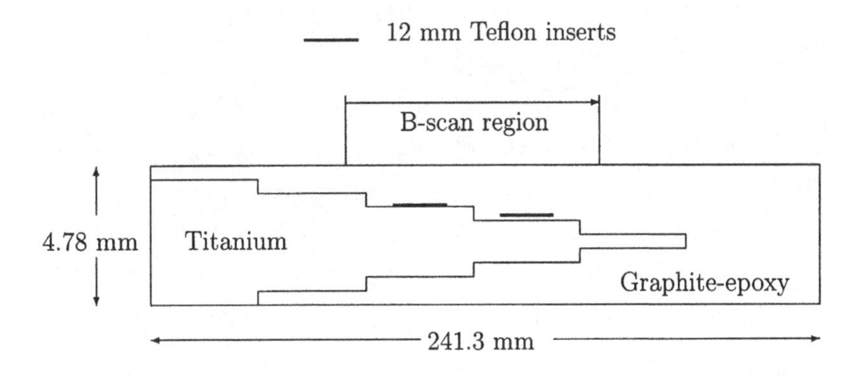


**Figure 2.3** Structural model of aircraft steplap joint specimen used for acquiring ultrasonic data.

by the wavelet can have several deleterious consequences. For example, the time location of the reflection may be ambiguous; also, two closely spaced reflections often cannot be resolved; finally, the polarity of the reflection (a useful indicator of a defect) may be obscured.

The temporal resolution of such data may be greatly improved by deconvolution, as shown in Figure 2.4. In this processing, each trace was deconvolved

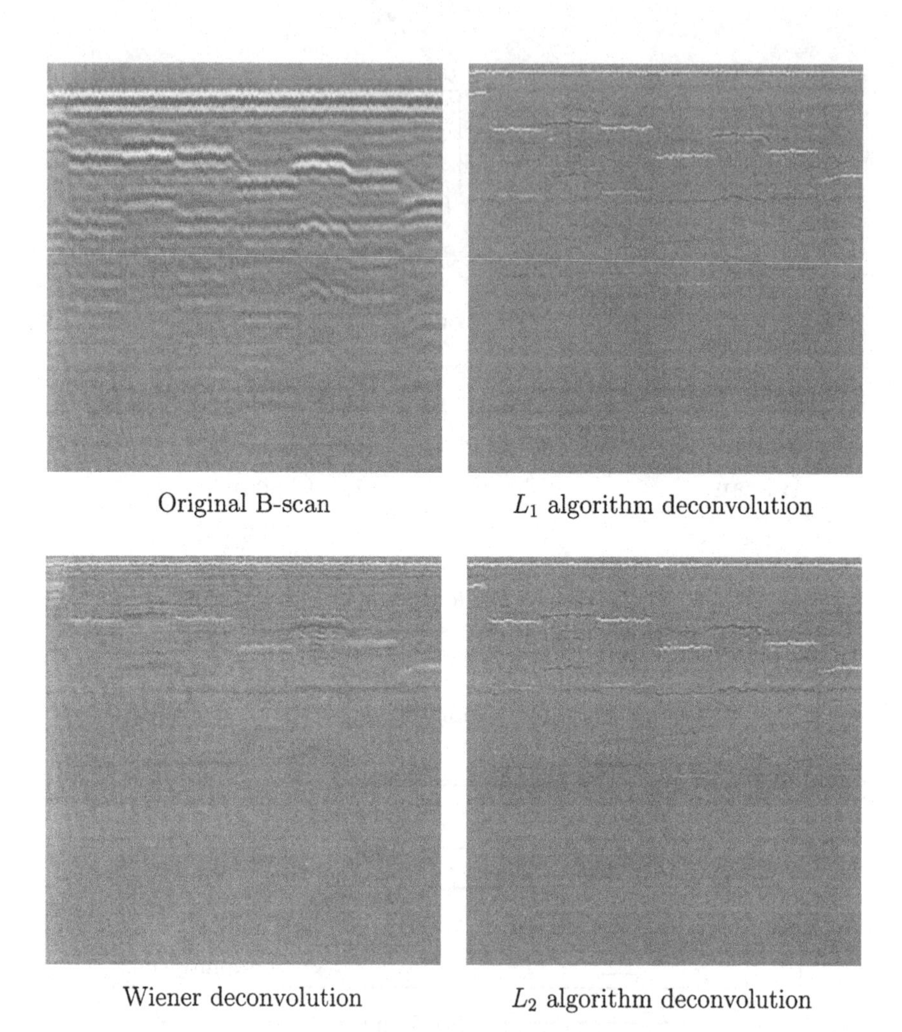


**Figure 2.4** Deconvolution processing of an ultrasonic B-scan. In each image, the horizontal dimension corresponds to the position of the transducer, while the vertical dimension corresponds to pressure as a function of time. Top left: original B-scan image, showing the spreading effect of the wavelet. Top right: image after deconvolution using the original $L_1$ deconvolution algorithm. Bottom left: image after linear (Wiener) deconvolution. Bottom right: image after deconvolution using the $L_2$ algorithm developed by Powell.

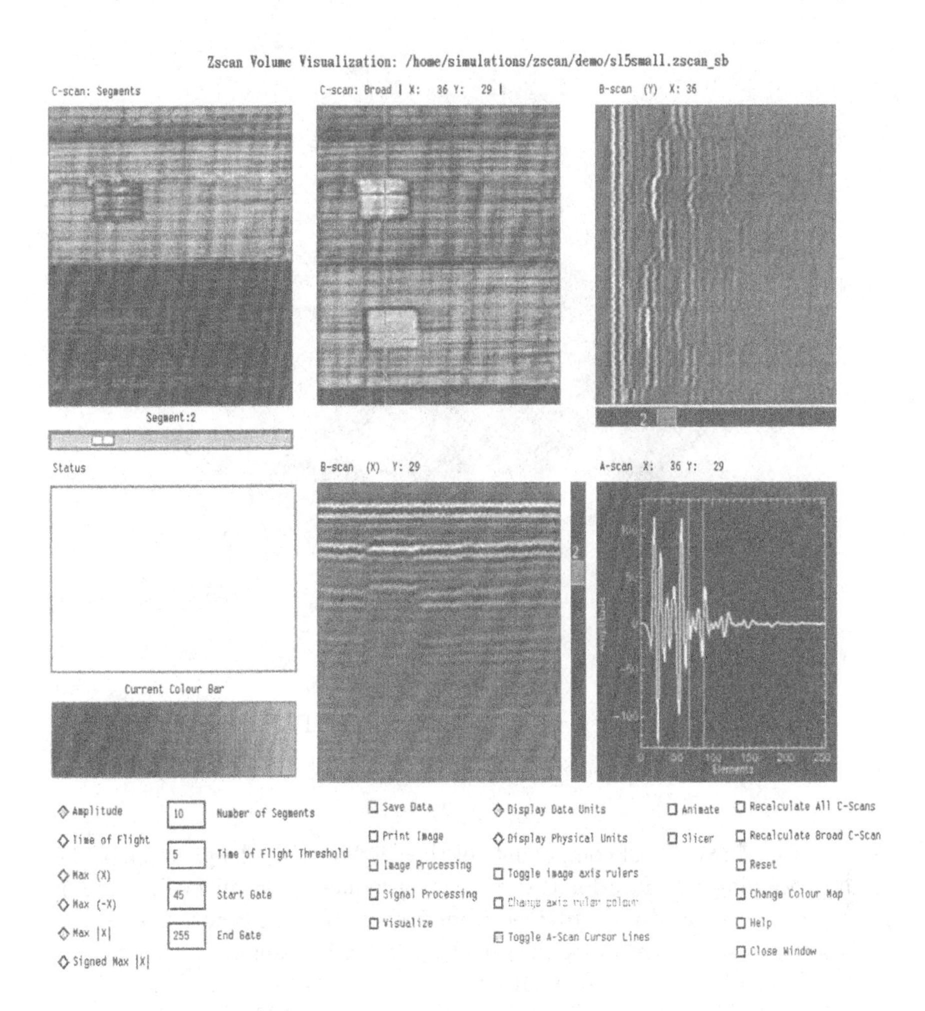


**Figure 2.5** Main display/processing screen of BCS's DETECT/ NDE System.

independently, using an estimate of the wavelet obtained by measuring the reflection of the transducer pulse from an isolated interface. The results of deconvolution using the above $L_1$ and $L_2$ algorithms, along with the result of conventional Wiener deconvolution, are displayed. It is evident that the $L_1$ and $L_2$ algorithms produce closely similar results, which are both more highly resolved than those obtained from Wiener deconvolution.

Powell's $L_2$ algorithm for sparse deconvolution has been incorporated as a module of a software package developed by BCS. This package, called DETECT/NDE, is designed for visualization and signal processing of large ultrasonic data sets. It is currently in use by the Canadian Department of National

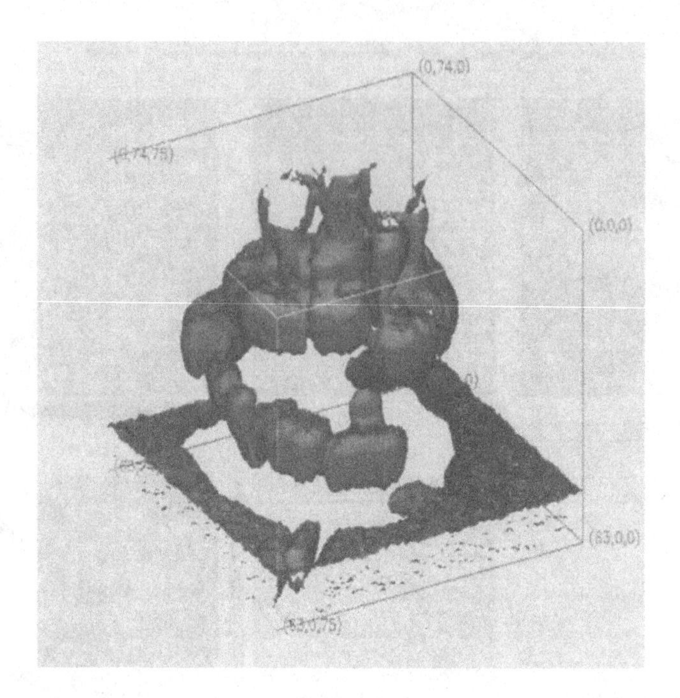


**Figure 2.6** 3-D view of spiral defect using BCS's DETECT/ NDE System.

Defence for inspection of components of their fighter jets. An example of the main processing interface for this software product is shown in Figure 2.5; the data set displayed was obtained from the specimen in Figure 2.3.

A second example of the ability of this system to visualize 3-D data sets is provided in Figure 2.6. The data set was obtained from a graphite-epoxy specimen which had been subjected to impact damage. The spatial extent of the damage and the spiral pattern of internal delaminations is clearly visible in this image.

In summary, the Powell sparse $L_2$ deconvolution algorithm has proved useful in several applications. However, there are still certain drawbacks to this and other sparse extraction algorithms. In particular, these approaches do not always give a good estimate of the solution when $\mathbf{W}$ is ill-conditioned (e.g., the bottom right estimated spike series in Figure 2.1). To provide more effective solutions to ill-posed deconvolution problems, BCS is currently investigating approaches based on extensions of Tikhonov regularization, and we are obtaining promising results in this area.

## 2.2 Inversion

Deconvolution is a special case of a more general class of inversion problems, in which the aim is to reconstruct an estimate of the acoustic impedance profile of a material. We have been concerned with methods of solving the 1-D inversion problem in geophysics and ultrasonic inspection for layered media in the presence of attenuation. In this problem, a series of $M$ discrete layers is assumed to model the data, with the $m$th layer characterized by an acoustic impedance $z_m$ for $m = 1, 2, \ldots, M$ and by an attenuation coefficient $\alpha_m$, along with a two-way travel time $h_m$ for $m = 1, 2, \ldots, M-1$. An $N$-element model trace $t(M, z, h, \alpha, w)$ is taken to be the convolution of a wavelet $w$ with an impulse response (spike series) $r(M, z, h, \alpha)$. An observed trace is represented by the sum of the model trace and noise. It should be noted that the estimation of the relative impedances of the layers is equivalent to the computation of reflection coefficients $c_m$ for the $M-1$ interfaces between the layers, through the relations

$$c_m = \frac{z_{m+1} - z_m}{z_{m+1} + z_m} \qquad \text{and} \qquad z_{m+1} = z_m \frac{1 + c_m}{1 - c_m}.$$

Consequently, given a set of impedances, the times and amplitudes of the reflections corresponding to the impulse response can be computed.

Frequently, especially in nondestructive evaluation, an initial model for the structure is available as prior knowledge (e.g., Figure 2.3 without the simulated defects). This may be used as a basis from which to detect defects, by modifying the structural model to account for the presence of certain types of defects, for those situations when the observed traces deviate significantly from those predicted by the initial model. For example, in composite materials, changes in the trace may result from changes in layer thickness, or from the presence of disbonds between two bonded sections of the materials, or from the presence of delaminations within the composite layer.

The problem may be stated as follows. Given a measured (data) trace $d$, known wavelet $w$ and layer attenuations $\alpha$, and initial estimates for $M, z$, and $h$, obtain final estimates of $M, z$, and $h$ which are consistent with the approximately known structure and above types of defects.

To solve this problem, an optimization procedure was adopted which minimizes

$$f(M, z, h) = \sum_{i=1}^{n} [d_i - t_i(M, z, h)]^2.$$

Two stages were involved in the approach: specification of the positions of the layers (corresponding to estimation of the travel times $h$), and computation of the impedances $z$ for each layer. Powell approached these problems by first developing a method for estimating $z$ for fixed $h$. This was implemented using the optimization package TOLMIN (also developed by Powell (1989)), and developing an algorithm for evaluating analytically the derivatives with respect to the impedances. The problem of optimizing the positions of the interfaces was then approached by allowing three types of layer manipulations whenever there was a significant discrepancy between the measured trace and the impedance-optimized trace. These manipulations were:

- adjustment of the thicknesses of each of the layers,
- addition of a new layer at a boundary (modelling a disbond),
- addition of a new layer entirely within another layer (modelling a delamination).

For each adjustment stage, the impedances were recomputed, and the residual error obtained. That set of manipulated layers which gave the minimum residual error was used as the estimate from which to model the structure of the material.

This approach to inversion was applied to the analysis of ultrasonic data by Zala and McRae (1991). In addition, the Powell $L_2$ deconvolution algorithm has also been applied to the related problem of high-resolution inversion in the absence of prior knowledge (Zala (1992)). In this application, it was used to identify the locations of the interfaces and to estimate the reflection coefficients.

# 3 Image warping using thin plate splines

In mine-hunting applications, side-scan sonar systems are used for navigation (route surveying) and analysis (mine detection and identification). One such system (shown in Figure 3.1) transmits two fan-shaped beams of ultrasonic energy from a towed sonar, which illuminate a strip of the sea floor via scan lines. As the towed sensor moves along its course, images are formed, based on the time delay and the intensity of signals reflected by the sea floor and any objects in the path of the beam. Images are geocoded by referencing raw data to a known ground system, using ship and towfish navigation data. Changes such as the presence of a new object are detected by comparing two geocoded images of the same scene.

Image comparison is often impaired by distortions arising from various causes, including sensor motion and sea floor irregularities. These distortions can hamper the detection of new objects, particularly smaller mine-like objects, but this can be largely overcome by the use of warping techniques. In this approach, the locations of features common to the two images are first identified. Warping functions are then defined which map features of one image (the sensed image) to corresponding features in the second image (the reference image). The remaining pixels in the sensed image are then warped by these same functions.

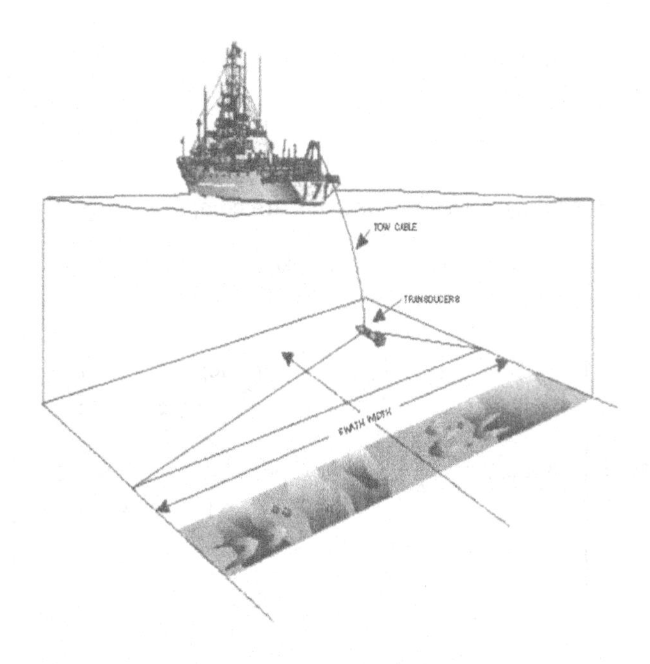


**Figure 3.1** A typical side-scan sonar system.

Thin plate splines (TPS's) are particularly suitable for this application, since, as solutions of the biharmonic equation, they have a minimum curvature property. For application to image warping, a separate TPS function is generated for shifting the $x$ and the $y$-components of the sensed image.

We consider two separate cases of TPS warping. In the first (simpler) case, the features to be mapped are points termed "control points". In the second, they may be represented as control points and curvilinear features; this application is termed "spider warping", since many of the features in our applications have a spidery resemblance.

## 3.1 Control point warping

Let $(x_i, y_i)$ represent the $i$th control point in the sensed image, and $(x'_i, y'_i)$ denote the corresponding control point in the reference image. In order to warp the $x$-components of all of the sensed image pixels based on exact interpolation at $n$ control points, we define a TPS warping function $w_x(x_i, y_i)$ with $(n + 3)$ parameters, satisfying

$$x'_i = w_x(x_i, y_i) = \sum_{j=1}^{n} \lambda_j r_i^2 \log r_i + a_0 + a_1 x_i + a_2 y_i,$$

where

$$r_i^2 = (x - x_i)^2 + (y - y_i)^2,$$

and the following conditions apply:

$$\sum_{j=1}^{n} \lambda_j = \sum_{j=1}^{n} \lambda_j x_j = \sum_{j=1}^{n} \lambda_j y_j = 0.$$

Writing $r_{ij} = r_i(x_j, y_j)$, the $\lambda_j$'s and $a_i$'s are computed from

$$\mathbf{Hv} = \mathbf{b},$$

where

$$\mathbf{H} = \begin{bmatrix} 0 & r_{21}^2 \log r_{21} & \dots & r_{n1}^2 \log r_{n1} & 1 & x_1 & y_1 \\ r_{12}^2 \log r_{12} & 0 & \dots & r_{n2}^2 \log r_{n2} & 1 & x_2 & y_2 \\ \vdots & \vdots & \ddots & \vdots & \vdots & \vdots & \vdots \\ r_{1n}^2 \log r_{1n} & r_{2n}^2 \log r_{2n} & \dots & 0 & 1 & x_n & y_n \\ 1 & 1 & \dots & 1 & 0 & 0 & 0 \\ x_1 & x_2 & \dots & x_n & 0 & 0 & 0 \\ y_1 & y_2 & \dots & y_n & 0 & 0 & 0 \end{bmatrix},$$

$$\mathbf{v} = \begin{bmatrix} \lambda_1 \\ \lambda_2 \\ \vdots \\ \lambda_n \\ a_0 \\ a_1 \\ a_2 \end{bmatrix}, \text{ and } \mathbf{b} = \begin{bmatrix} x_1' \\ x_2' \\ \vdots \\ x_n' \\ 0 \\ 0 \\ 0 \end{bmatrix}.$$

Note that the $(n+3) \times (n+3)$ matrix $\mathbf{H}$ depends only on the sensed data.

A similar TPS function $w_y(x_i, y_i)$ is defined for the $y$-components of the control points:

$$y_i' = w_y(x_i, y_i) = \sum_{j=1}^{n} \mu_j r_i^2 \log r_i + b_0 + b_1 x_i + b_2 y_i,$$

and the determination of these parameters involves the same matrix $\mathbf{H}$.

In practice, application of TPS warping to image registration problems typically involves hundreds or even thousands of control points. Moreover, image warping requires the subsequent evaluation of two TPS's at each pixel, which may number in the millions. Until recently, a severe drawback of this application was the computational effort required to use the TPS model. Powell has made contributions in both problem areas: he developed algorithms for solving the interpolation equations efficiently (Powell (1995a)), and he designed a fast scheme which tabulates TPS values on a very fine rectangular grid (Powell (1992)). The latter has resulted in an improvement of speed for typical applications by a factor of more than a hundred, thereby bringing TPS warping techniques into the realm of practical applicability.

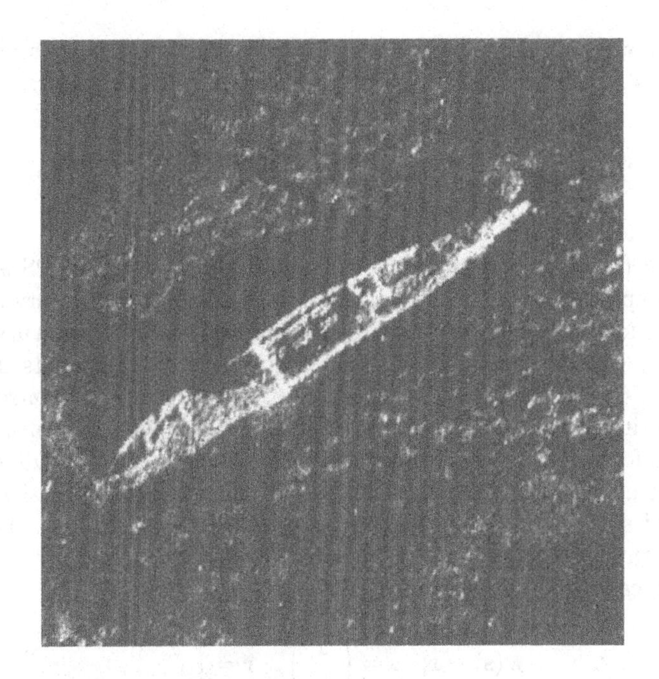

**Figure 3.2** Side-scan sonar image of a 55-m wreck.

A TPS-warped version of the wreck image in Figure 3.2 is provided in Barrodale *et al.* (1993), together with further details. BCS has incorporated a TPS control point warping module into an extensive Sonar Image Processing System (SIPS), currently being developed for the Esquimalt Defence Research Detachment in Victoria. This SIPS, which employs an object relational database containing sonar images, is designed to support new Canadian mine-hunting technology using ships and remotely controlled vehicles.

## 3.2 Spider warping

In the spider warping application, the aim is to extend TPS warping to curvilinear features in addition to point features. In this way, distributed features in the sensed image may be warped into corresponding features in the reference image.

As before, define $\mathbf{w}(\mathbf{z}): R^2 \to R^2$ as

$$\mathbf{w}(\mathbf{z}) = \left\{ \begin{array}{l} \sum_{j=1}^{n} \lambda_j \phi(||\mathbf{z}-\mathbf{z}_j||) + p(\mathbf{z}) \\ \sum_{j=1}^{n} \mu_j \phi(||\mathbf{z}-\mathbf{z}_j||) + q(\mathbf{z}) \end{array} \right\}, \qquad \mathbf{z} \in R^2,$$

where $\phi(r) = r^2 \log r$, $r > 0$, and $\phi(0) = 0$, and $p$ and $q$ are linear polynomials from $R^2$ to $R$. If the sensed image includes $n$ points $\mathbf{s}_i$, which have to map into $n$ points $\mathbf{r}_i$ in the reference image, then $\mathbf{w}$ has to satisfy $\mathbf{w}(\mathbf{s}_i) = \mathbf{r}_i$, $i = 1, \ldots, n$. These interpolation equations, together with the constraints

$$\sum_{j=1}^{n} \lambda_j = \sum_{j=1}^{n} \mu_j = 0 \text{ and } \sum_{j=1}^{n} \lambda_j \mathbf{s}_j = \sum_{j=1}^{n} \mu_j \mathbf{s}_j = \mathbf{0}$$

uniquely determine the $\lambda_j$'s, $\mu_j$'s, $p$ and $q$.

Powell (1995b), in response to our requests, developed a TPS algorithm which maps control points to control points *and* prescribed curves to prescribed curves ("spider legs"). In this approach, each chosen curve in the sensed image is first replaced by a finite sequence of discrete points. Let these points be $\{\mathbf{s}_i : i = m+1, \ldots, n\}$, where $m$ is the number of control points. The algorithm chooses the position of the points $\mathbf{r}_i = \mathbf{w}(\mathbf{s}_i)$, $i = m+1, \ldots, n$, on the reference image curves automatically. This is done by adjusting the positions of $\mathbf{r}_i$, $i = m+1, \ldots, n$ in order to minimize the integral $I$, defined below, of squares of second derivatives of the TPS function $\mathbf{w}$. Each trial value of this integral requires $O((n-m)^2)$ computer operations.

Specifically, for the transformation

$$\mathbf{w}(\mathbf{s}) = \mathbf{r}, \quad \mathbf{s} = \begin{pmatrix} x \\ y \end{pmatrix}, \quad \mathbf{r} = \begin{pmatrix} x' \\ y' \end{pmatrix},$$

the integral $I$ is defined as

$$I \equiv \int_{R^2} \left\{ \left(\frac{\partial^2 x'}{\partial x^2}\right)^2 + 2\left(\frac{\partial^2 x'}{\partial x \partial y}\right)^2 + \left(\frac{\partial^2 x'}{\partial y^2}\right)^2 + \left(\frac{\partial^2 y'}{\partial x^2}\right)^2 + 2\left(\frac{\partial^2 y'}{\partial x \partial y}\right)^2 + \left(\frac{\partial^2 y'}{\partial y^2}\right)^2 \right\} dx dy.$$

It is well known that among all functions satisfying the $n$ interpolation equations $\mathbf{w}(\mathbf{s}_i) = \mathbf{r}_i$, $I$ is a minimum when $\mathbf{w}$ is a TPS.

Let $H$ be the coefficient matrix for these equations, as given in Section 3.1, and let $A$ be the leading $n \times n$ submatrix of $H^{-1}$. It can be shown that $I$ is a convex, quadratic function of the reference data, which can be expressed as

$$I = 16\pi \sum_{j=1}^{n} \sum_{k=1}^{n} (x'_j a_{jk} x'_k + y'_j a_{jk} y'_k).$$

Note that the elements $a_{jk}$ of $A$ depend only on sensed data, which, in the case of spider warping, include $m$ control points and $(n-m)$ spider points (discretized curves). The $m$ sensed control points are mapped exactly onto the reference control points, whereas the $(n-m)$ sensed spider points are mapped to reference curves, each of which is represented in parametric form as $(x'(\theta), y'(\theta))$.

The optimization problem is to find values of the variables $\{\theta_i : i = 1, 2, \ldots, n - m\}$ that minimize $I$, while preserving the sequences of spider points as they are mapped from sensed spider legs to reference spider legs.

At the time of writing, our computational experience with spider warping has been restricted to model problems with $n \leq 1,000$; the Powell algorithm has been very effective in these cases. BCS is currently involved in two environmental projects in which TPS surfaces interpolating 100,000 points may prove useful, and spider warping with $n \geq 10,000$ may be attempted. We note, with anticipation and optimism, that Powell's current research includes development of algorithms for problems of large size!

# References

[1] Barrodale, I., R. Kuwahara, R. Poeckert and D. Skea (1993) "Side-scan sonar image processing using thin plate splines and control point matching", *Numerical Algorithms* **5**, 85–98.

[2] Chapman, N.R. and I. Barrodale (1983) "Deconvolution of marine seismic data using the $L_1$ norm", *Geophys. J. Roy. Astr. Soc.* **72**, 93–100.

[3] Claerbout, J.F. and F. Muir (1973) "Robust modelling with erratic data", *Geophysics* **38**, 826–844.

[4] Milinazzo, F., C. Zala and I. Barrodale (1987) "On the rate of growth of condition numbers of convolution matrices", *IEEE Tr. Acoust. Sp. Sig. Proc.* **ASSP-35**, 471–475.

[5] Powell, M.J.D. (1988) private communication.

[6] Powell, M.J.D. (1989) "TOLMIN: A Fortran package for linearly constrained optimization calculations," DAMTP Report 1989/NA2, University of Cambridge.

[7] Powell, M.J.D. (1992) "Tabulation of thin plate splines on a very fine two-dimensional grid," in *Numerical Methods in Approximation Theory,* **9** (D. Braess and L. Schumaker, eds), Birkhäuser Verlag (Basel), 221–244.

[8] Powell, M.J.D. (1995a) "A new iterative algorithm for thin plate spline interpolation in two dimensions," DAMTP Report 1995/NA10, University of Cambridge.

[9] Powell, M.J.D. (1995b) "A thin plate spline method for mapping curves into curves in two dimensions," DAMTP Report 1995/NA02, University of Cambridge.

[10] Taylor, H.L., S.C. Banks and J.F. McCoy (1979) "Deconvolution with the $L_1$ norm", *Geophysics* **44**, 39–52.

[11] Zala, C.A. and K.I. McRae (1991) "An optimization method for acoustic impedance estimation of layered structures using prior knowledge", in *Acoustical Imaging* (Vol. 18), (H. Lee and G. Wade, eds), Plenum Press (New York), 363–372.

[12] Zala, C.A. (1992) "High-resolution inversion of ultrasonic traces", *IEEE Trans. Ultrason., Ferroelec., Freq. Contr.* **UFFC 39**, 458–463.

# On the Meir/Sharma/Hall/Meyer Analysis of the Spline Interpolation Error

*Carl de Boor*

Center for Mathematical Sciences, University of Wisconsin, Madison, WI 53705, USA[1]

**Abstract**

The use of B-splines is shown to simplify existing arguments (such as those by Sharma and Meir, Hall, Hall and Meyer, Howell and Varma, and others) concerning the error in certain univariate spline interpolation which rely on the diagonal dominance of certain tridiagonal matrices. In addition, a simple recipe for the calculation of the constant in the sharp error bound is obtained.

## 1 The story

The basic idea of the Meir/Sharma/Hall/Meyer error analysis for cubic spline interpolation (see [14], [5], [8]) has been extended to cover also certain deficient quartic and quintic spline interpolation schemes, the latter already by them, the former by Howell and Varma [9] and, perhaps, others. It is the purpose of this note to describe the most general situation to which this idea applies and thereby, perhaps, to obviate further papers on various special cases.

In full generality, the idea covers the following situation. Let

$$\tau = (\tau_1, \dots, \tau_{k+1}) = (\underbrace{0, \dots, 0}_{\rho \text{ times}} < \tau_{\rho+1} \leq \cdots \leq \tau_{\rho+r} < \underbrace{1, \dots, 1}_{\rho \text{ times}}) \tag{1.1}$$

be any nondecreasing $(k+1)$-sequence in $[0\,..\,1]$ in which both 0 and 1 occur exactly $\rho > 0$ times, hence $r$ is such that

$$k + 1 = 2\rho + r.$$

Denote by

$$H_\tau g$$

[1]This work was supported by the National Science Foundation under Grants DMS-9224748 and DMS-9626319, and by the U.S. Army Research Office under Contract DAAH04-95-1-0089.

the Hermite interpolant to $g$ at $\tau$, i.e., the unique polynomial of degree $\le k$ which agrees with $g$ at $\tau$, repeated points in $\tau$ corresponding to the matching of derivative values in the standard way.

The corresponding Hermite spline interpolant, for a given break sequence

$$\xi = (a = \xi_1 < \cdots < \xi_{\ell+1} = b)$$

in the interval $[a \mathinner{..} b]$, is, by definition, the function

$$Hg := H_{\tau,\xi} g := \sigma_i^{-1} H_\tau \sigma_i g \quad \text{on} \quad [\xi_i \mathinner{..} \xi_{i+1}], \quad i = 1, \ldots, \ell, \tag{1.2}$$

with $\sigma_i$ the linear change of variables

$$\sigma_i f : t \mapsto f(\xi_i + t\, h_i), \qquad h_i := \xi_{i+1} - \xi_i, \quad i = 1, \ldots, \ell.$$

By the Schoenberg–Whitney theorem, there exists, for given smooth $g$, exactly one element in the space

$$S := \Pi_{k,\xi}^{(\rho)}$$

of pp functions in $C^{(\rho)}[a \mathinner{..} b]$ of degree $k$ with break sequence $\xi$ which matches $g$ $(\rho - 2)$-fold at each breakpoint, except at $a$ and $b$ where the match is $(\rho - 1)$-fold, and matches also at the points

$$\xi_i + \tau_\kappa h_i, \quad \kappa = \rho + 1, \ldots, \rho + r, \quad i = 1 \ldots, \ell.$$

We denote this element by

$$s = Ig = I_{\tau,\xi} g.$$

Special cases include:

(i) $k = 3$, $\rho = 2$, hence $r = 0$, leading to *complete cubic spline interpolation.*

(ii) $k = 3$, $\rho = 1$, hence $r = 2$, leading to *deficient cubic spline interpolation* studied by Dikshit and Powar [4] (and others).

(iii) $k = 2$, $\rho = 1$, hence $r = 1$, leading to *parabolic spline interpolation* (at midpoints, e.g.), first discussed by Subbotin [15].

(iv) $k = 4$, $\rho = 2$, hence $r = 1$, leading to $C^2$-*quartic spline interpolation*, as investigated by Howell and Varma [9].

(v) $k = 5$, $\rho = 3$, hence $r = 0$, leading to $C^3$-*quintic spline interpolation*, as investigated by Sharma and Meir [14], as well as by Hall [5], and Hall and Meyer [8].

The *construction* of $Ig$ proceeds as follows. Since each polynomial piece of $Ig$ is completely determined once the numbers $D^{\rho-1}Ig(\xi_i)$, $i = 2, \ldots, \ell$, are known, and thereby joins its neighboring piece(s) in $C^{(\rho-1)}$-fashion, one relies on the tridiagonal linear system

$$\mathrm{A}(D^{\rho-1}Ig) = \Lambda g \tag{1.3}$$

obtained from the requirement that the various polynomial pieces should join in $C^{(\rho)}$-fashion. Somewhat more explicitly,

$$(\mathrm{A}f)(i) = \sum_j A(i,j)f(\xi_j), \qquad i = 2, \ldots, \ell$$

for a certain tridiagonal *matrix* $A$, with $A(i,j)$ expressible in terms of $h_{i-1}, h_i$, $k$, and $\rho$ (see (3.1)). Also,

$$\Lambda g = (\lambda_i g : i = 2, \ldots, \ell),$$

with $\lambda_i g$ a certain linear combination of the data

$$g|_{\tau^{(i)}}$$

(see (1.8)) for $g$ on the interval $[\xi_{i-1} \mathinner{..} \xi_{i+1}]$, all $i$. It is one of the results of the present note (see (3.2)) that, as has been known in all the special cases mentioned above, the matrix $A$ of the linear system (1.3) is diagonally dominant, uniformly in $\xi$ if scaled appropriately and if $\tau$ is symmetric, hence the linear system can be solved stably by Gauss elimination without pivoting.

The Meir/Sharma/Hall/Meyer *analysis of the error* $g - Ig$ is based on the split

$$e := g - Ig = (g - Hg) + (H - I)g,$$

as first used by Hall [5] (for cubic and quintic spline interpolation).

The first part is dealt with locally, relying on known sharp bounds (such as those in Birkhoff and Priver [2] for the cubic and quintic case) for Hermite interpolation. Fortunately, Shadrin [13] quite recently gave a complete treatment of such sharp error bounds, for arbitrary degree, thus providing a proof that

$$\|g - Hg\|_{\infty,(\xi_i..\xi_{i+1})} \le c_\tau h_i^{k+1} \|D^{k+1}g\|_{\infty,(\xi_i..\xi_{i+1})},$$

with

$$c_\tau := \|(1 - H_\tau)(\cdot)^{k+1}\|_\infty/(k+1)!.$$

This implies that

$$\|g - Hg\|_\infty \le c_\tau (\max_i h_i)^{k+1} \|D^{k+1}g\|_\infty \tag{1.4}$$

with equality for any $g$ for which $D^{k+1}g$ is absolutely constant and is constant in each break interval $(\xi_i \mathinner{..} \xi_{i+1})$.

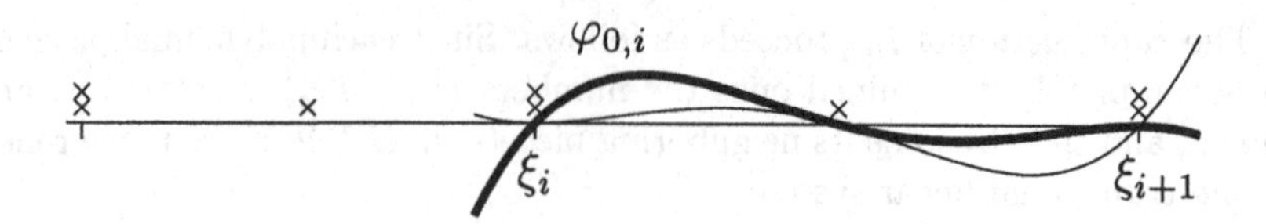


**Figure 1.1** The sequence $\tau^{(i)}$ ($\times$) and the polynomials $\varphi_{0,i}$ and $\varphi_{1,i}$ for $\rho = 2$, $r = 1$, hence $k = 4$.

The second part requires comparison of the (local) Hermite interpolant, $Hg$, with the (global) spline interpolant, $s = Ig$. On each break interval $[\xi_i \,.\,.\, \xi_{i+1}]$, both are polynomials of degree $\le k$ that match the same information about $g$ *except* for the $(\rho - 1)$st derivative at the endpoints which is only matched by $Hg$, while $Ig$ gets that information from the linear system (1.3). It follows that, on this break interval,

$$Hg - Ig = D^{\rho-1}e(\xi_i)\varphi_{0,i} + D^{\rho-1}e(\xi_{i+1})\varphi_{1,i},$$

with $\varphi_{j,i} := h_i^{\rho-1}\sigma_i\varphi_j$, $j = 0, 1$, and $\varphi_0, \varphi_1$ certain polynomials of degree $\le k$ with

$$(-1)^{k-\rho}\varphi_0\varphi_1 \ge 0 \quad \text{on } [0\,.\,.\,1]. \tag{1.5}$$

Therefore,

$$\|Hg - Ig\|_{\infty,(\xi_i..\xi_{i+1})} \le \|D^{\rho-1}e|_\xi\|_\infty c_\varphi h_i^{\rho-1},$$

with

$$c_\varphi := \|\varphi_0 + (-1)^{k-\rho}\varphi_1\|_\infty,$$

and with equality if $D^{\rho-1}e|_\xi$ is absolutely constant and

$$(-1)^{k-\rho}D^{\rho-1}e(\xi_i)D^{\rho-1}e(\xi_{i+1}) \ge 0$$

for all $i$. Further, the sequence $(D^{\rho-1}e(\xi_j) : j = 2, \ldots, \ell)$ satisfies the linear system (1.3), but with an appropriately changed right side:

$$\mathrm{A}(D^{\rho-1}e) = \mathrm{A}(D^{\rho-1}g) - \Lambda g =: \mathrm{M}g. \tag{1.6}$$

The coefficient matrix, $A$, for this linear system is tridiagonal, with positive diagonal entries. Hence, if

$$c_A := \min_i(A(i,i) - |A(i,i-1)| - |A(i,i+1)|) \tag{1.7}$$

is positive, then

$$\|(D^{\rho-1}e)|_\xi\|_\infty \le \|\mathrm{M}g\|_\infty / c_A.$$

Further, if $\tau$ is symmetric, i.e., $\tau = 1 - \tau$, then (see (3.2)) the minimum in (1.7) is taken on for $i = 3, \ldots, \ell - 1$ and is independent of $\xi$. Finally,

$$\mathrm{M}g = (\mu_i g : i = 2, \ldots, \ell),$$

with each $\mu_i$ a certain linear combination of the linear functionals used for $\sigma_\nu^{-1} H_\tau \sigma_\nu$, $\nu = i-1, i$, i.e., $\mu_i g$ is a linear combination of values and derivatives of $g$ at the entries of

$$\tau^{(i)} := (\xi_{i-1} + \tau h_{i-1}\,,\ \xi_i + (\tau_\kappa : \kappa = \rho+1, \ldots, k+1) h_i\,). \tag{1.8}$$

In the early papers, local Taylor expansion was appealed to to assert (sometimes incorrectly, e.g., in [5] for the cubic case) that $\mu_i g = c_i D^{k+1} g(\eta_i)$ for some $\eta_i \in (\xi_{i-1} \mathinner{\ldotp\ldotp} \xi_{i+1})$. In any case, $\mu_i$ must vanish on $\Pi_k$, hence, as first used explicitly in Howell and Varma [9] (but see already Schultz [12]), for smooth $g$,

$$\mu_i g = \int_{\xi_{i-1}}^{\xi_{i+1}} \widehat{\mu}_i D^{k+1} g, \quad i = 2, \ldots, \ell,$$

for a certain pp function $\widehat{\mu}_i$, with breaks only at the data points and with support in $[\xi_{i-1} \mathinner{\ldotp\ldotp} \xi_{i+1}]$. Therefore,

$$\|\mathrm{M} g\|_\infty \le \max_i \|\widehat{\mu}_i\|_1 \, \|D^{k+1} g\|_\infty,$$

and a detailed analysis (see (4.2)) of the integrals

$$\|\widehat{\mu}_i\|_1 = \int_{\xi_{i-1}}^{\xi_{i+1}} |\widehat{\mu}_i(y)| \mathrm{d}y$$

provides the formula

$$\|\widehat{\mu}_i\|_1 \;=\; c_\mu \frac{h_i h_{i-1}^{k-\rho+2} + h_{i-1} h_i^{k-\rho+2}}{h_{i-1} + h_i} \tag{1.9}$$

with a certain $\xi$-independent constant $c_\mu$, and so finishes the derivation of the error bound

$$\|g - Ig\|_\infty \le (c_\tau + (c_\mu / c_A) c_\varphi)(\max_i h_i)^{k+1} \|D^{k+1} g\|_\infty. \tag{1.10}$$

Curtis and Powell [3] seem to be the first to have proved sharpness of spline interpolation error bounds, for the special case of cubic spline interpolation on a uniform mesh. Their argument relies on a detailed analysis of the Peano kernel for the error.

In the present setting, Hall and Meyer [8] were the first to consider the sharpness of the bound (1.10) (and of corresponding bounds on the error in derivatives). To be precise, this sharpness is only asymptotic, as $\ell$ grows large, with the error bound almost exact for a uniform $\xi$. In all cases considered, i.e., Hall and Meyer in cubic and $C^{(3)}$-quintic spline interpolation, Howell and Varma in $C^{(2)}$-quartic spline interpolation, the sharpness has been established by exhibiting an extremizing function, in effect an Euler spline in the cases considered by Hall and Meyer, and the monomial $(\,\cdot\,)^{k+1}$ in $C^{(2)}$-quartic spline interpolation. The argument for it relies on an often quite detailed analysis of the function $\widehat{\mu}_i$ to show its simple sign pattern (just one sign change, at $\xi_i$, in the cubic and quintic case, no sign change in the quartic case) and on a further, ad hoc, argument involving the matrix $A$.

It is one purpose of this note to point out that the error bound (1.10) is asymptotically sharp in the general case, and that this can be seen quite easily. In particular, $\widehat{\mu}_i$ is necessarily a linear combination of the $k+1-\rho$ B-splines of order $k+1$ associated with the knot sequence $\tau^{(i)}$ (see (1.8)) and has a zero of order $k-\rho$ at $\xi_i$, and this already determines it uniquely, up to sign. In particular, $\widehat{\mu}_i$ is of one sign on each of the two breakpoint intervals in its support. Furthermore, the $\widehat{\mu}_i$ are of one sign (resp., change sign) exactly when the inverse of $A$ is of one sign (resp., checkerboard), making it possible to exhibit a function for which the error bound is (asymptotically) sharp. In particular, the error bound (1.10) can thereby be seen to be exact in a very simple case, in which the error can be computed directly, thus making the separate calculation of the four constants, $c_\tau$, $c_\mu$, $c_A$, and $c_\varphi$ unnecessary.

Here is the formal statement.

**Theorem 1** *Let $\tau = (\tau_i : i = 1, \ldots, k+1)$ be a nondecreasing sequence in $[0\,..\,1]$ in which both 0 and 1 occur exactly $\rho > 0$ times, and set $r := k+1-2\rho$ (hence $0 < \tau_{\rho+1} \leq \cdots \leq \tau_{\rho+r} < 1$). Let $\xi = (a = \xi_1 < \cdots < \xi_{\ell+1} = b)$ and let $S = \Pi^{(\rho)}_{k,\xi}$ be the space of pp functions in $C^{(\rho)}$ of degree $k$ with breaks $\xi$. Then, for every $g \in C^{(\rho-1)}[a\,..\,b]$, there exists exactly one $s = Ig$ in $S$ which interpolates to $g$ in the sense that*

$$s = g \ \text{ at } \xi_j + (\tau_2, \ldots, \tau_k)\Delta\xi_j, \quad j = 1, \ldots, \ell$$

*(with repetitions indicating the matching of derivatives in the usual way), and also*

$$D^{\rho-1}s = D^{\rho-1}g \ \text{ at } a, b. \tag{1.11}$$

*Further, if $\tau$ is symmetric, i.e., $\tau = 1 - \tau$, then*

$$\|g - Ig\|_\infty \leq c\,(\max_i \Delta\xi_i)^{k+1}\|D^{k+1}g\|_\infty, \tag{1.12}$$

*with $c := \|g_0 - I_0 g_0\|_\infty$ the maximum error in the special case that $\xi = (-1, 0, 1)$, $g_0$ is 2-periodic with $D^{k+1}g_0(x)$ equal to $-1$ for $-1 < x < 0$ and equal to $(-1)^{k-\rho-1}$ for $0 < x < 1$, and with (1.11) replaced by the periodic end conditions*

$$D^{\rho-j}s(b) = D^{\rho-j}s(a), \quad j = 0, 1. \tag{1.13}$$

*In particular (1.12) is asymptotically sharp, for a uniform $\xi$, as $\ell \to \infty$.*

It should be noted that the symmetry assumption for $\tau$ is a convenience. However, diagonal dominance cannot be had without *some* assumption on $\tau$.

Also, it should be pointed out that sharp *pointwise* error bounds are available for interpolation by splines with simple (interior) knots to data at simple data sites which satisfy the Schoenberg–Whitney conditions; see Sections 5.2

and 5.3 of Korneichuk's book [10]. Those results can easily be extended to the present situation (which involves non-simple knots and data sites). This note's virtue (if any) lies in pointing out a simple way to derive the error bound (1.12) and compute the exact constant $c$ in it for the particular spline interpolation schemes considered.

The remainder of this note proves the various assertions made, thus providing all missing details for the proof of Theorem 1.

## 2 Some facts concerning $H_\tau$

In this preparatory section, we derive various facts concerning the Hermite interpolant, $H_\tau g$, in particular its dependence on the data $D^{\rho-1}g(\nu)$, $\nu = 0, 1$, of use later, in the analysis of the matrix of the linear system (1.3) and of the norms $\|\widehat{\mu}_i\|_1$.

Since $H_\tau g$ matches $g$ at $\tau$, we may write it in the convenient form

$$H_\tau g = D^{\rho-1}g(0)\varphi_0 + D^{\rho-1}g(1)\varphi_1 + Q_\tau g, \tag{2.1}$$

with

$$\varphi_\nu = \alpha_\nu(\cdot + \nu - 1)\psi, \quad \nu = 0, 1,$$

where

$$\psi(t) := \prod_{\kappa=2}^{k}(t - \tau_\kappa),$$

and the constant $\alpha_\nu$ is such that $D^{\rho-1}\varphi_\nu(\nu) = 1$, $\nu = 0, 1$, hence $D^{\rho-1}Q_\tau g(\nu) = 0$, $\nu = 0, 1$, and $Q_\tau g$ depends only on the data $g|_{(\tau_\kappa:\kappa=2,\ldots,k)}$. In other words, (2.1) gives $H_\tau g$ in 'Lagrange form', but with only two of the data, namely $D^{\rho-1}g$ at 0 and at 1, mentioned explicitly, and the rest of the information collected in the term $Q_\tau g$.

Here are some details concerning the function $\varphi_1$ which, together with the analogous information about $\varphi_0$, will be needed in the discussion of the diagonal dominance of $A$ and the sign pattern of its inverse. Since

$$D^q(\cdot - \beta)\psi = qD^{q-1}\psi + (\cdot - \beta)D^q\psi, \quad q = 0, 1, 2, \ldots,$$

we have

$$1 = D^{\rho-1}\varphi_1(1) = \alpha_1(\rho - 1)D^{\rho-2}\psi(1) + \alpha_1 D^{\rho-1}\psi(1).$$

Since

$$D^{\rho-2}\psi(1) = 0 \neq D^{\rho-1}\psi(1),$$

this implies that $\alpha_1 = 1/D^{\rho-1}\psi(1)$, hence

$$D^\rho\varphi_1 = \left(\rho D^{\rho-1}\psi + (\cdot)^1 D^\rho\psi\right)/D^{\rho-1}\psi(1).$$

In particular,

$$D^\rho\varphi_1(0) = \rho D^{\rho-1}\psi(0)/D^{\rho-1}\psi(1),$$

while

$$D^\rho\varphi_1(1) = \rho + D^\rho\psi(1)/D^{\rho-1}\psi(1) \;>\; \rho \tag{2.2}$$

(since, by Rolle's theorem, $D^{\rho-1}\psi$ has all its zeros in the open interval $(0\,..\,1)$).

Since $D^{\rho-1}\psi$ has exactly $\rho - 1 + r = k - \rho$ zeros in $(0\,..\,1)$, we have

$$(-1)^{k-\rho}D^{\rho-1}\psi(0)D^{\rho-1}\psi(1) \;>\; 0,$$

hence

$$(-1)^{k-\rho}D^\rho\varphi_1(0) \;>\; 0\,. \tag{2.3}$$

This implies that $(-1)^{k-\rho}D^{2\rho-1}(\varphi_0\varphi_1)(0) > 0$, hence, since $\varphi_0\varphi_1$ vanishes to exact order $2\rho - 1$ at 0 and has only even zeros in $(0\,..\,1)$, (1.5) follows. More explicitly than (2.3),

$$D^\rho\varphi_1(0) = (-1)^{k-\rho}\rho, \quad \text{if } \tau \text{ is symmetric.}$$

It seems most efficient to deduce the corresponding statements for $\varphi_0$ from the fact that

$$\varphi_0 = \varphi_{0,\tau} = (-1)^{\rho-1}\varphi_{1,1-\tau}(1 - \cdot).$$

This implies that

$$D^\rho\varphi_0 = (-\rho\, D^{\rho-1}\psi + (1 - \cdot)D^\rho\psi)/D^{\rho-1}\psi(0)$$

and, in particular,

$$D^\rho\varphi_0(0) = -\rho + D^\rho\psi(0)/D^{\rho-1}\psi(0) \;<\; -\rho$$

(since $D^{\rho-1}\psi$ has all its zeros in the open interval $(0\,..\,1)$). Therefore,

$$D^\rho\varphi_0(0) \;< 0 \;< D^\rho\varphi_1(1). \tag{2.4}$$

# 3 The tridiagonal linear system

Since the Schoenberg–Whitney theorem guarantees existence and uniqueness of the interpolant, we know that the coefficient matrix $A$ in (1.3) is invertible. We now show that, for a symmetric $\tau$, $A$ is diagonally dominant, independently of $\xi$, with positive diagonal entries and with the next-to-diagonal entries negative (positive) exactly when $k - \rho$ is even (odd), hence the inverse of $A$ is positive (checkerboard).

The $i$th equation in (1.3) expresses the requirement that the interpolant, $s = Ig$, have a continuous $\rho$th derivative at $\xi_i$:

$$D^\rho s(\xi_i-) = D^\rho s(\xi_i+).$$

Written out in more detail, this reads, after reordering so as to put the unknown terms on the left and the given information on the right (except that, for $i = 2$ and $i = \ell$, we leave the known endpoint derivative on the left side),

$$\begin{aligned}&(D^{\rho-1}s(\xi_{i-1})D^\rho\varphi_0(1) + D^{\rho-1}s(\xi_i)D^\rho\varphi_1(1))/h_{i-1}\\ &\qquad -\\ &(D^{\rho-1}s(\xi_{i+1})D^\rho\varphi_1(0) + D^{\rho-1}s(\xi_i)D^\rho\varphi_0(0))/h_i\\ &= D^{\rho-1}(\sigma_{i-1}^{-1}Q_\tau\sigma_{i-1}g)(\xi_i-) - D^{\rho-1}(\sigma_i^{-1}Q_\tau\sigma_i g)(\xi_i+).\end{aligned}$$

After multiplying both sides by $(h_{i-1}h_i)/(h_{i-1}+h_i)$, we obtain the $i$th equation of (1.3), with

$$A(i,j) = \frac{1}{h_{i-1}+h_i}\begin{cases} D^\rho\varphi_0(1)h_i, & j = i-1;\\ D^\rho\varphi_1(1)h_i - D^\rho\varphi_0(0)h_{i-1}, & j = i;\\ -D^\rho\varphi_1(0)h_{i-1}, & j = i+1.\end{cases} \tag{3.1}$$

By (2.4), $A(i,i) > 0$ while, by (2.3), $A(i,i\pm1)$ is negative (positive) exactly when $k-\rho$ is even (odd).

If now $\tau$ is symmetric, i.e., $\tau = 1-\tau$, then

$$A(i,j) = \frac{1}{h_i+h_{i-1}}\begin{cases} -\rho(-1)^{k-\rho}h_i, & j = i-1;\\ (\rho + D^\rho\psi(1)/D^{\rho-1}\psi(1))(h_i+h_{i-1}), & j = i;\\ -\rho(-1)^{k-\rho}h_{i-1}, & j = i+1.\end{cases} \tag{3.2}$$

Therefore,

$$A(i,i) - \sum_{j\neq i}|A(i,j)| \geq D^\rho\psi(1)/D^{\rho-1}\psi(1) =: c_A, \tag{3.3}$$

with $c_A$ positive (by (2.2)) and with equality for all $i$ except for the first and last, unless we switch to the periodic end conditions (1.13), in which case there is equality here for all $i$.

While it is trivial that, therefore, $A$ is also diagonally dominant for all 'nearby' $\tau$, $A$ is not diagonally dominant for all choices of $\tau$. E.g., for the simplest possible case $\rho = 1 = r$, hence $k = 2$, we have $\phi_0 = (\cdot - \tau_2)(\cdot - 1)/\tau_2$, and therefore

$$\begin{aligned}&A(i,i) - |A(i,i-1)| - |A(i,i+1)|\\ &= \left[h_i\left(\frac{2-\tau_2}{1-\tau_2} - \frac{1-\tau_2}{\tau_2}\right) + h_{i-1}\left(\frac{\tau_2+1}{\tau_2} - \frac{\tau_2}{1-\tau_2}\right)\right]/(h_i+h_{i-1}),\end{aligned}$$

which, for some $\xi$, becomes negative when $\tau_2 \notin [2-\sqrt{3}\,..\,1/\sqrt{2}]$. (However, this matrix is *column* diagonally dominant for all $\tau_2 \in (0\,..\,1)$).

# 4 A representer for $\mu_i$

Since the error, $e = g - Ig$, is zero for any $g \in S$ and the matrix $A$ is invertible, it follows that, for each $i$, the functional $\mu_i$ necessarily vanishes on $S$. This implies that

$$\mu_i g = \int_{\xi_{i-1}}^{\xi_{i+1}} \widehat{\mu}_i D^{k+1} g, \quad i = 2, \ldots, \ell,$$

for a certain pp function $\widehat{\mu}_i$, with breaks only at the data sites and with support in $[\xi_{i-1} \,..\, \xi_{i+1}]$. Precisely, any smooth $g$ can be written

$$g(t) = \sum_{j \le k} D^j g(a)(t-a)^j/j! + \int_a^b (t-y)_+^k/k! \; D^{k+1} g(y) \mathrm{d}t,$$

and, since $\Pi_k|_{[a..b]} \subset S|_{[a..b]}$, this implies that

$$\widehat{\mu}_i(y) = \mu_i(\cdot - y)_+^k/k! \;=\; -\mu_i(\cdot - y)_-^k/k!,$$

the second equality since $(\cdot - y)_+^k + (\cdot - y)_-^k \in \Pi_k$. Consequently, $\widehat{\mu}_i$ is an element of the space $S_{k+1,\tau^{(i)}}$ of splines of order $k+1$ with knot sequence $\tau^{(i)}$ (see (1.8)). Any spline space has dimension equal to (number of knots) − (order), hence,

$$\dim S_{k+1,\tau^{(i)}} = \rho + r = k - \rho + 1. \tag{4.1}$$

Since also $(\cdot - \xi_i)_+^q$, $q = \rho+1, \ldots, k$, is in $S$, hence is annihilated by $\mu_i$, it follows that $\widehat{\mu}_i$ vanishes $(k-\rho)$-fold at $\xi_i$. It follows from (4.1) (and, e.g., the Schoenberg–Whitney theorem) that this condition alone determines $\widehat{\mu}_i$ uniquely, up to a scalar factor. Further, since $\widehat{\mu}_i \ne 0$, it follows that $\widehat{\mu}_i$ vanishes in $(\xi_{i-1} \,..\, \xi_{i+1})$ only at $\xi_i$, hence is of one sign, or changes sign only at $\xi_i$, depending on whether $k - \rho$ is even or odd.

This simple observation suffices for the derivation of an explicit formula for $\|\mu_i\| = \|\widehat{\mu}_i\|_1$, as follows. Consider the functions

$$\varphi_\pm(t) := (t - \beta_\pm)\varphi_0\left(\frac{t - \xi_i}{h_\pm}\right)(h_\pm)^k \gamma_\pm D^{\rho-1}\psi(0)/(k+1)!,$$

with $\gamma_- := -1$, $\gamma_+ := -(-1)^{k-\rho}$, $h_\pm := \xi_{i\pm1} - \xi_i$, with $\varphi_0$ and $\psi$ as in the detailed discussion of $H_\tau$, and with $\beta_\pm$ to be determined in such a way that

$$D^{\rho-j}\varphi_-(\xi_i) \;=\; D^{\rho-j}\varphi_+(\xi_i), \quad j = 0, 1.$$

It turns out that $\beta_\pm$ can so be determined. The resulting function

$$g(t) := \begin{cases} \varphi_-(t) & t \le \xi_i; \\ \varphi_+(t) & t \ge \xi_i, \end{cases}$$

is seen to be piecewise polynomial of degree $k+1$, with $D^{k+1}g$ equal to $\gamma_-$ on $(\xi_{i-1} \,..\, \xi_i)$, and equal to $\gamma_+$ on $(\xi_i \,..\, \xi_{i+1})$, hence

$$\int_{\xi_{i-1}}^{\xi_{i+1}} \widehat{\mu}_i D^{k+1} g \;=\; \pm\|\widehat{\mu}_i\|_1.$$

On the other hand, $g|_{\tau^{(i)}} = 0$, and even $D^{\rho-1}g(\xi_{i\pm1}) = 0$. This implies that the spline interpolant to $g$ for the break sequence $(\xi_{i-1}, \xi_i, \xi_{i+1})$ is zero, therefore $A(i,i)D^{\rho-1}g(\xi_i) = \int_{\xi_{i-1}}^{\xi_{i+1}} \widehat{\mu}_i D^{k+1}g$.

This gives (1.9) in the more explicit form

$$\|\widehat{\mu}_i\|_1 = \frac{\rho|D^{\rho-1}\psi(0)|}{(k+1)!} \frac{h_i h_{i-1}^{k-\rho+2} + h_{i-1}h_i^{k-\rho+2}}{h_i + h_{i-1}}, \tag{4.2}$$

and even shows that $\widehat{\mu}_i(t) \le 0$ for $t \le \xi_i$, while $(-1)^{k-\rho}\widehat{\mu}_i(t) \le 0$ for $t \ge \xi_i$ (by virtue of the fact that $A(i,i)D^{\rho-1}g(\xi_i)$ is positive). To be sure, the $g$ constructed is only in $C^{(\rho)}$. However, it is the error in the spline interpolant to any $(k+1)$st primitive of the function $D^{k+1}g$, and that is all that really matters.

A similar construction for the simple break sequence $(-1, 0, 1)$ can be used to provide the 2-periodic function $g_0$, with

$$g_0(x) = (x(x-1) + \rho D^{\rho-1}\psi(0)/D^{\rho}\psi(0))\psi(x)/(k+1)!$$

on $(0\,..\,1)$ and odd (even) when $\rho - 1$ is odd (even), for which $I_0 g_0 = 0$ while $\|g_0\|_\infty = c_\tau + (c_\mu/c_A)c_\varphi$.

It also follows that, if $k - \rho$ is even, then all the $\mu_i$ take on their norm on the function $(\cdot)^{k+1}$, while, if $k - \rho$ is odd, they all take on their norm on the function whose $(k+1)$st derivative is absolutely constant and changes sign across each interior $\xi_i$. In the former case, the corresponding right side in (1.6) is of one sign, while, in the latter case, it is maximally alternating in sign. In the latter case, and for a symmetric choice of $\tau$ and for a uniform $\xi$, it follows that $D^{\rho-1}e$ vanishes at all the breaks $\xi_i$ when $g = (\cdot)^{k+1}$, hence the error in $D^{\rho-1}e$ at the breaks is at least one order higher than expected. This was first observed for cubic spline interpolation, in [1].

Now recall that, correspondingly, the inverse of the matrix $A$ in (1.6) is of one sign when $k-\rho$ is even, and is checkerboard when $k-\rho$ is odd. Therefore, (1.12) is asymptotically sharp for a uniform $\xi$ (using the essentially local character of spline interpolation in that case).

# References


[1] Birkhoff, G. and C.R. de Boor (1965) "Piecewise polynomial interpolation and approximation" in *Approximation of Functions* (H.L. Garabedian, ed.), Elsevier (New York), 164–190.

[2] Birkhoff, G. and A.S. Priver (1967) "Hermite interpolation errors for derivatives", *J. Math. Phys.* **46**, 440–447.

[3] Curtis, A.R. and M.J.D. Powell (1967) "Error analysis for equal-interval interpolation by cubic splines", Theoret. Phys. Div. A.E.R.E., Technical report R.5600.

[4] Dikshit, H.P. and P.L. Powar (1981) "On deficient cubic spline interpolation", *J. Approx. Theory* **31**, 99–106.

[5] Hall, C.A. (1968) "On error bounds for spline interpolation", *J. Approx. Theory* **1**, 209–218.

[6] Hall, C.A. (1969) "Error bounds for periodic quintic splines", *CACM* **12**, 450–452.

[7] Hall, C.A. (1973) "Uniform convergence of cubic spline interpolants", *J. Approx. Theory* **7**, 71–75.

[8] Hall, C.A. and W.W. Meyer (1976) "Optimal error bounds for cubic spline interpolation", *J. Approx. Theory* **16**, 105–122.

[9] Howell, G. and A.K. Varma (1989) "Best error bounds for quartic spline interpolation", *J. Approx. Theory* **58**, 58–67.

[10] Korneichuk, N. (1991) *Exact Constants in Approximation Theory*, Encyclopedia of Mathematics and its Applications 38, Cambridge University Press (Cambridge, UK).

[11] Rana, S.S. and Y.P. Dubey (1996) "Local behaviour of the deficient discrete cubic spline interpolator", *J. Approx. Theory* **86(1)**, 120–127.

[12] Schultz, M.H. (1973) *Spline Analysis*, Prentice–Hall (Englewood Cliffs, New Jersey).

[13] Shadrin, A.Yu. (1995) "Error bounds for Lagrange interpolation", *J. Approx. Theory* **80**, 25–49.

[14] Sharma, A. and A. Meir, "Degree of approximation of spline interpolation", *J. Math. Mech.* **15(5)**, 759–767.

[15] Subbotin, Y.N. (1967) "On piecewise-polynomial (spline) interpolation", *Mat. Zametki* **1**, 63–70.

# Asymptotically Optimal Approximation and Numerical Solutions of Differential Equations

*Martin D. Buhmann*

Departement of Mathematics, ETH Zentrum, 8092 Zürich, Switzerland

*Charles A. Micchelli*

Mathematical Sciences Department, IBM Thomas J. Watson Research Center, Yorktown Heights, NY 10598, USA

*Amos Ron*

Department of Computer Science, University of Wisconsin–Madison, 1210 West Dayton Street, Madison, WI 43706, USA[1]

**Abstract**

Given a finite subset $J \in \mathbb{R}^n$ and a point $\lambda \in \mathbb{R}^n$, we study in this paper the possible convergence, as $h \to 0$, of the coefficients in least-squares approximations to $f(\cdot + \lambda h)$ from the space spanned by $(f(\cdot + jh))_{j \in J}$. We invoke the 'least solution' of a polynomial interpolation problem to show that the coefficients converge for a generic $J$ and $\lambda$, provided the underlying function $f$ is sufficiently smooth. Moreover, in certain cases (such as when $J \cup \lambda$ form a rectangular mesh), the limits of the least-squares coefficients are shown to be independent of $f$ and are characterised by their polynomial accuracy. Finally, we employ a different argument to show that convergence of the least-squares coefficients also occurs for a certain class of functions which are not necessarily smooth.

The above study is highly relevant to the problem of selecting an optimal finite difference scheme for solving partial differential equations, a connection that is also discussed in the paper.

[1]Amos Ron's work was supported by the National Science Foundation under Grants DMS-9224748 and DMS-9626313, and by the U.S. Army Research Office under Contract DAAH04-95-1-0089.

# 1 Introduction

An optimal finite difference method for the numerical solution of the Cauchy problem for a given partial differential equation is, by definition, the scheme that minimises the local truncation error after one step. In this paper we study certain extremal problems that are closely related to optimal finite difference schemes for finding numerical solutions of such problems. For relevant general information on difference methods connected with the Cauchy problem see, e.g., (Iserles and Strang, 1983).

Consider now the concrete problem of finding an optimal finite difference method for approximating the solution of a well-posed Cauchy problem using the linear constant-coefficient differential equation

$$\begin{aligned} u_t(x,t) &= p(D_x)u(x,t), \quad t \geq 0, \\ u(x,0) &= f(x), \qquad x \in \mathbb{R}, \end{aligned}$$

where $p$ is a polynomial and $D_x = \frac{\partial}{\partial x}$. It is shown in (Micchelli and Miranker, 1973a, 1973b) that the optimal method is computable in terms of the optimal finite difference method for the advection partial differential equation

$$u_t = u_x, \qquad t \geq 0, \quad x \in \mathbb{R},$$

whose solution is $u(x,t) = f(x+t)$ with the appropriate ranges of $x$ and $t$. If $h$ denotes the mesh increment in the $x$-direction and $\lambda h$ is the mesh increment in the $t$ direction, $\lambda$ being the Courant number, then

$$f(x+\lambda h) - \sum_{j\in J} c_j f(x+jh), \qquad x \in \mathbb{R}, \tag{1.1}$$

is the local truncation error after one time step corresponding to a generic finite difference scheme that uses the pointset $J$ at the backward level of time. This means that our goal can be accomplished by solving a corresponding least-squares minimisation problem which is the univariate case ($n = 1$) of Problem A below. In the statement of the problem, we use the notation $E^t$ for the shift operator, i.e.

$$E^t : f \mapsto f(\cdot + t). \tag{1.2}$$

**Problem A** *Given $f \in L^2(\mathbb{R}^n)$, a finite set $J \subset \mathbb{R}^n$, a point $\lambda \in \mathbb{R}^n$, and $h > 0$, minimise the error function*

$$\|E^{h\lambda} f - \sum_{j\in J} c_j E^{hj} f\|_{L^2(\mathbb{R}^n)} \tag{1.3}$$

*over all $c := (c_j)_{j\in J} \in \mathbb{C}^J$. The optimal coefficients are denoted here and hereafter by $c(\lambda, h) := (c_j(\lambda, h))_{j\in J}$.*

We can therefore view the stated problem as the problem of approximating the $h\lambda$-translate of some given function $f$ from the finite-dimensional space

$$S_h^J(f) := \mathrm{span}\{E^{jh}f \mid j \in J\},$$

$J$ being a given finite subset of $\mathbb{R}^n$. We think of the function $f$ as given to us (the initial value of the above partial differential equation); on the other hand, the choice of the set $J$ and the scaling parameter $h$ are within our discretion.

Since the space $S_h^J(f)$ is finite dimensional, it is seemingly feasible to find the exact solution to Problem A, i.e., the least-squares approximation to $E^{h\lambda}f$ from $S_h^J(f)$. This, however, requires the recomputation of the approximant whenever $f$ or $h$ is changed. Furthermore, since the least-squares solution is expressed via the basis $\{E^{jh}f \mid j \in J\}$ of $S_h^J(f)$ (i.e. we compute the coefficients with respect to that basis), the deteriorating condition number of that basis, as $h \to 0$, may lead to loss of significance through numerical instability of the process. In addition, knowing the theoretical background of such an approximation should help us in determining the size of $h$ to be used, and the configuration and cardinality of the set $J$ to be chosen. A large set $J$ will make the computation of the least-squares solution computationally prohibitive, while a small set $J$ may force us to select a scaling parameter $h$ that is too small.

This problem has already been dealt with in the literature if $f \in L^1(\mathbb{R}) \cap L^2(\mathbb{R})$ and for a positive $\alpha$

$$\lim_{x\to\pm\infty} |\hat{f}(x)||x|^{\alpha+\frac{1}{2}} = \text{const} \neq 0 \tag{1.4}$$

in the above-mentioned two papers by Micchelli and Miranker. Specifically, it is proved there that, upon assuming (1.4), the optimal coefficients $c_j(\lambda, h)$ converge, for every $j \in J$, to a limit $c_j(\lambda) := \lim_{h\to 0} c_j(\lambda, h)$; that limit is referred to as the *principal part* of the intermediate coefficient $c_j(\lambda, h)$. This allows the $h$-independent principal parts to be employed as a suitable alternative to the optimal coefficients in (1.1). Moreover, it is the principal part of the optimal scheme which determines its stability.

Although then the local truncation error is not necessarily minimal, we will show that it is 'near optimal' in a certain sense, and it allows for a simpler, easier to compute finite difference scheme. In other words, while the coefficients of the best approximation to $E^{h\lambda}f$ from $S_h^J(f)$ may exhibit undesired numerical behaviour as $h \to 0$, one explicitly identifies cases where the coefficients of the best approximation each converge to a limit (defined as the principal part of the coefficient), and considers the approximation scheme when the $c_j(\lambda, h)$ are replaced by their principal parts.

In extending these univariate results we have two approaches in this paper. One involves a multivariate generalisation of (1.4) which is also more general even in one dimension. It leads to so-called radial basis function interpolation, and the principal parts turn out to be Lagrange functions for this interpolation with respect to the set $J$.

Another approach involves multivariate polynomial interpolation on the set $J$. Indeed, when we use this second *ansatz*, the message of this paper concerning Problem A becomes strikingly simple, and so we are going to describe it first. The statements below require $J$ to be in a *total degree* configuration. This notion is defined near the end of Section 2. Here, we mention that any pointset $J$ in $\mathbb{R}$ satisfies the total degree assumption, and that, in two dimensions, 3-sets $J$ whose points are not collinear, as well as 6-sets $J$ whose points do not lie on a conic, are of total degree. The *degree* $m(J)$ of a total degree configuration $J$ is also defined in Section 2. We remark here that $m(J) = |J| - 1$ in case $J \subset \mathbb{R}$, and that the degrees of the two-dimensional 3-sets and 6-sets discussed above are 1 and 2 respectively.

One of the main results of this paper, that invokes the polynomial interpolation approach, is as follows. In its statement, and elsewhere in the paper, we use the notation $e_j$, $j \in \mathbb{R}^n$, for the exponential with frequency $j$,

$$e_j : t \mapsto e^{ij \cdot t},$$

and

$$\Pi_m$$

for the space of all $n$-variate polynomials of degree $\leq m$.

**Theorem 1** *Let $J$ be a total degree subset of $\mathbb{R}^n$ of degree $m(J)$. Then, for every $\lambda \in \mathbb{R}^n$, there is a sequence $c(\lambda) \in \mathbb{C}^J$, that depends only on $\lambda$ and $J$, with the following properties:*

*(1) For every $\alpha > m(J)$, and for every function in the Sobolev space $W_2^\alpha(\mathbb{R}^n)$, the optimal coefficients $c_j(\lambda, h)$, $j \in J$, of Problem A converge each to $c_j(\lambda)$.*

*(2) For every $\alpha \leq m(J) + 1$, and for every $f \in W_2^\alpha(\mathbb{R}^n)$, the $L^2(\mathbb{R}^n)$-error in the approximation scheme*

$$E^{h\lambda} f \approx \sum_{j \in J} c_j(\lambda) E^{hj} f$$

*is $O(h^\alpha)$ as $h \to 0$.*

*Furthermore, the sequence $c(\lambda)$ is characterised by its polynomial accuracy: it is the unique sequence supported on $J$ that satisfies*

$$p(\lambda) = \sum_{j \in J} c_j(\lambda) p(j), \quad \forall p \in \Pi_{m(J)}.$$

The analysis that leads to Theorem 1 is given in Sections 3 and 4. We mention that the convergence asserted in (1) is valid for configurations $J$ that are more general than the 'total degree' mentioned here; however, in these more general setups, either the limit coefficients are not universal, i.e., they may depend on $f$, and/or the convergence is proved for only certain, specific $\lambda$. For example, convergence to universal limits occurs if $J \cup \lambda$ consists of the vertices of a rectangular mesh.

Thus, for a large collection of functions and for quite general configurations $J$, the optimal coefficients of Problem A converge to an $f$-independent sequence $c(\lambda)$. Further, that sequence 'works well' for other functions too (even though for these other functions the optimal coefficients may not converge at all), in the sense that the error $\|E^{h\lambda} - \sum_{j \in J} c_j(\lambda) e_j\|_{L^2(\mathbb{R}^n)}$ decays at rates that are related to the smoothness of $f$. This leads us naturally to examine also $f$-independent $h$-independent schemes of the form

$$E^{h\lambda} \approx \sum_{j \in J} c_j E^{hj} f.$$

We call schemes such as the above 'near optimal' if they provide approximation rates like those in (2) of Theorem 1.

The paper is laid out as follows. In Section 2, we review relevant facts concerning the *least solution of the polynomial interpolation problem* (de Boor and Ron, 1990). That least solution turns out to be the main tool in our study of Problem A for 'sufficiently smooth' functions $f$ that is carried out in Section 4. In that section, Theorem 1 is proved, together with some more general results. Before, in Section 3, we study the problem of near optimal schemes. In Section 5, we consider the convergence of the optimal coefficients in case the underlying function $f$ is not sufficiently smooth for the application of Theorem 1. We identify situations when the Fourier transform of $f$ is asymptotically homogeneous at $\infty$ (similarly to (1.4)) and prove that the optimal coefficients converge, regardless of the configuration of $J$ and/or the choice of $\lambda$. As stated before, the limit coefficients are identified as the Lagrange functions of certain interpolation problems that involve radial basis functions (Buhmann, 1993, Dyn, 1989, Micchelli, 1986).

# 2 The least solution of the polynomial interpolation problem

We briefly discuss here the least solution of the polynomial interpolation problem of (de Boor and Ron, 1990), a tool that we require in our analysis in Section 4, and to a lesser extent in our analysis of Section 3.

Given a finite $J \subset \mathbb{R}^n$, and a polynomial space $P$ in $n$ variables, we say that $P$ **is correct** for interpolation on $J$ if every data given on $J$ can be interpolated by a unique $p \in P$. For each $J$, there are of course many polynomial spaces $P$ that interpolate correctly on $J$. In (de Boor and Ron, 1990, 1992a, 1992b) a correct polynomial space $\Pi_J$, the least solution, of least possible degree and various other desired properties was constructed. That construction applies to any $J$, and is fitted to the problem we discuss in this paper: the limit of the optimal coefficients (the principal parts) for sufficiently smooth functions $f$ will be expressed in terms of this least solution. We review below some of its basic features and refer the reader to the above-mentioned references for further discussions.

Let $\mathrm{Exp}_J$ be the exponential space

$$\mathrm{Exp}_J := \mathrm{span}\{e_j \mid j \in J\}.$$

Since each $g \in \mathrm{Exp}_J$ is entire, it can be written uniquely as a convergent sum

$$g = g_0 + g_1 + g_2 + \ldots,$$

with each $g_m$ a *homogeneous* polynomial of degree $m$. One sets

$$g_\downarrow$$

(read '$g$ least') to be the non-zero polynomial $g_m$ of *least* degree in the above expansion. The **least solution** of the polynomial interpolation problem is then the *homogeneous* polynomial space

$$\Pi_J := \mathrm{span}\{g_\downarrow \mid g \in \mathrm{Exp}_J\}.$$

The space $\Pi_J$ is correct for interpolation on $J$. In fact, it is of minimal degree among all such polynomial spaces: if there exists any polynomial space $P$ that is correct for $J$ and contains the space $\Pi_\alpha$ (of all polynomials of degree $\leq \alpha$), then $\Pi_J$ contains $\Pi_\alpha$, as well. Also, $\Pi_J$ coincides with standard choices of correct spaces, in cases such choices exist, for example when $J$ forms a rectangular grid.

We now associate with the set $J$ two important parameters: its *minimal degree* (or *accuracy*) $m(J)$, and its *maximal degree* $M(J)$; both notions are essential in our analysis of Problem A.

**Definition 1** *Let $J$ be a finite subset of $\mathbb{R}^n$, $\Pi_J$ its associated least polynomial space.*

*(1) $m(J)$ is the maximal integer $m$ for which $\Pi_m \subset \Pi_J$.*

*(2) $M(J)$ is the minimal $m$ for which $\Pi_J \subset \Pi_m$.*

EXAMPLE In one dimension, the least solution of any $m$-set $J$ is the space $\Pi_{m-1}$. Thus, we always have $m(J) = M(J) = |J| - 1$. In more than one dimension, an equality $m(J) = M(J) =: m$ implies that $\Pi_J = \Pi_m$, and is possible only in case the cardinality $|J|$ matches the dimension of $\Pi_\alpha$ for some $\alpha$. However, with $k := \dim \Pi_m$, for some $m$, the equality $m(J) = M(J) = m$ holds for a generic $k$-set in $\mathbb{R}^n$.

The above example motivates the following definition:

**Definition 2** *We say that* $J \subset \mathbb{R}^n$ *is*

*(1) in* **general position** *if* $M(J) \leq m(J) + 1$ *and*

*(2) of* **total degree** *if* $m(J) = M(J)$.

We note that 'general position' is the generic case. However, various important configurations for $J$, such as the vertices of a rectangular grid, are not in general position. 'Total degree' occurs when $J$ is in general position, and further its cardinality matches the dimension of some $\Pi_m$; for example, in two dimensions, three points that are not collinear, and six points that do not lie on a conic are of total degree. In contrast, a set consisting of four or five points (still in $\mathbb{R}^2$) cannot be of total degree. Such a set will be in general position, though, unless all its points lie on one line.

# 3 Near optimality

We say that '$S_h^J()$ provides approximation order $\alpha$ to $f \in L^2(\mathbb{R}^n)$', if $\operatorname{dist}(E^{h\lambda}f, S_h^J(f)) = O(h^\alpha)$, for every/some $\lambda \in \mathbb{R}^n$. The range of relevant $\lambda$ will be clear in each context. Of course, the optimal coefficients of Problem A realise any approximation order that can be provided (after all, they are the best). However, our objective in this paper is to replace the $h$-dependent optimal coefficients by $h$-independent ones. Therefore, it seems useful to consider the following problem:

**Problem B** *Assume that* $S_h^J()$ *provides, for some fixed* $J$ *and* $\lambda$, *approximation order* $\alpha > 0$ *to all functions in some smoothness class* $L$. *Are there* $h$-*independent sequences* $c \in \mathbb{C}^J$ *that realise that order, i.e., such that*

$$\|E^{h\lambda}f - \sum_{j \in J} c_j E^{hj} f\|_{L^2(\mathbb{R}^n)} = O(h^\alpha), \quad \forall f \in L\,?$$

Schemes that use such sequences $c$ as described above are called **near optimal** because they realise the approximation rate $\alpha$ of best approximation, but not necessarily with the same constant factor.

One of the fundamental principles of Approximation Theory is the intimate relation between the decay rate of the error in approximation schemes and the smoothness class of the function being approximated. Roughly speaking, one expects that 'functions with only $\alpha$ derivatives' will be approximated at rates no better than $\alpha$. This also explains the custom of studying rates of convergence simultaneously for all functions in the same smoothness class. For our particular problem, the close relation between the decay of the error with $h \to 0$ and the smoothness class of $f$ is even more basic: the mere definition of 'smoothness' via the moduli of smoothness notion (see, e.g., DeVore and Lorentz, 1993, p. 44) shows that smoothness can be defined, hence interpreted, as the ability to approximate a function well by nearby translates of it.

Theorem 1 provides simultaneous answers to both Problems A and B: it shows that for smooth functions the optimal coefficients of Problem A converge to their universal limits and that for functions of lesser smoothness these universal coefficients provide the expected approximation orders. This result is stated with respect to Sobolev spaces. However, it is still valid if we replace these spaces by the larger *Besov spaces* that we now define. Let

$$\Omega_j := \{t \in \mathbb{R}^n \mid 2^{j-1} \leq \|t\| < 2^j\}$$

and

$$\Omega_0 := \{t \in \mathbb{R}^n \mid \|t\| < 1\},$$

where $\|\cdot\|$ denotes the Euclidean norm on $\mathbb{R}^n$. For any $f \in L^2(\mathbb{R}^n)$ we let $a_j(f) := \|\|\cdot\|^\alpha \hat{f}\|_{L^2(\Omega_j)}$. The **Besov space** $B^\alpha_\infty(L^2(\mathbb{R}^n))$ is the space that contains all $f$ with uniformly bounded $a_j(f)$:

$$\|f\|_{B^\alpha_\infty(L^2(\mathbb{R}^n))} := \sup_{j\geq 0} \|\,\|\cdot\|^\alpha \hat{f}\|_{L^2(\Omega_j)}.$$

**Theorem 2** *Let $J \subset \mathbb{R}^n$ be given, let $\lambda \in \mathbb{R}^n$, and $c = (c_j)_{j\in J} \in \mathbb{C}^J$. Suppose that, with*

$$E_c := e_\lambda - \sum_{j\in J} c_j e_j,$$

*$|E_c(t)| = O(\|t\|^m)$ as $\|t\| \to 0$ for some positive integer $m$. Then, with*

$$A_h f := \sum_{j\in J} c_j E^{hj} f:$$

*(1) For every $f \in W_2^m(\mathbb{R}^n)$, and $h \to 0$,*

$$\|E^{h\lambda} f - A_h f\|_{L^2(\mathbb{R}^n)} = O(h^m).$$

*(2) For every $\alpha < m$, every $f \in B^\alpha_\infty(L^2(\mathbb{R}^n))$, and $h \to 0$,*

$$\|E^{h\lambda} f - A_h f\|_{L^2(\mathbb{R}^n)} = O(h^\alpha).$$

**Proof** Measuring the error in the Fourier domain, one has

$$\|E^{h\lambda}f - A_h f\|_{L^2(\mathbb{R}^n)} = \tfrac{1}{(2\pi)^n}\|\widehat{f}E_c(h\cdot)\|_{L^2(\mathbb{R}^n)} \leq \tfrac{1}{(2\pi)^n}\|(h\|\cdot\|)^m\widehat{f}\|_{L^2(\mathbb{R}^n)} = \|f\|_{W_2^m(\mathbb{R}^n)}h^m,$$

which proves (1). For the proof of (2), we assume without loss of generality that $h = 2^{-k}$ for some $k$. We estimate

$$\int |E_c(ht)|^2|\hat{f}(t)|^2\,dt. \tag{3.1}$$

We divide the integral into its range of integration over $\|t\| > 2^k = h^{-1}$ and the remaining part. We deal with the former part first. Here, since $f \in B^\alpha_\infty(L^2(\mathbb{R}^n))$, we may write, for a generic positive constant $C$,

$$\int_{\Omega_j} |\hat{f}(t)|^2\,dt \leq 2^{-2(j-1)\alpha}\int_{\Omega_j}\|t\|^{2\alpha}|\hat{f}(t)|^2\,dt \leq C2^{-2j\alpha}.$$

Summing over $j \geq k+1$, we obtain that

$$\int_{\|t\|>2^k}|E_c(t)|^2|\hat{f}(t)|^2\,dt \leq C\int_{\|t\|>2^k}|\hat{f}(t)|^2\,dt \leq C2^{-2k\alpha} = O(h^{2\alpha}).$$

For $j \leq k$, we have

$$\begin{aligned}\int_{\Omega_j}|E_c(t/2^k)|^2|\hat{f}(t)|^2\,dt &\leq C2^{-2km}\int_{\Omega_j}\|t\|^{2m}|\hat{f}(t)|^2\,dt \\ &\leq C2^{-2km}2^{2j(m-\alpha)}\int_{\Omega_j}\|t\|^{2\alpha}|\hat{f}(t)|^2\,dt.\end{aligned}$$

Invoking the fact that $f \in B^\alpha_\infty(L^2(\mathbb{R}^n))$, we can sum over $j = 0, 1, \ldots, k$, to obtain the required $O(h^{2\alpha})$ bound. □

Discussion Expressed with the least term of a smooth function (as defined in Section 2), the requirement in Theorem 2 concerning $E_c$ is that $\deg E_{c\downarrow} \geq m$. If $m \leq m(J)+1$, with $m(J)$ the accuracy of $J$, then, for each $\lambda \in \mathbb{R}^J$, this requirement can be fulfilled by a suitable choice of $c$: for any $J \subset \mathbb{R}^n$, and any polynomial $p$ of degree at most $m(J)$, there exists $g \in \mathrm{Exp}_J$ whose Taylor expansion up to degree $m(J)$ yields $p$. For higher values of $m$, such a condition may be satisfied only for very special values of $\lambda$ that lie in the zero sets of certain polynomials. In any event, if $m > M(J)+1$, then the aforementioned condition cannot hold for any $\lambda \notin J$.

**Proposition 1** *Given $J \subset \mathbb{R}^n$, the spaces $(S^J_h())_h$ provide approximation order*

*(1) $m(J)+1$ to the Sobolev space $W_2^{m(J)+1}(\mathbb{R}^n)$ and*

*(2) for $\alpha < m(J)+1$, $\alpha$ to the Besov space $B^\alpha_\infty(L^2(\mathbb{R}^n))$.*

The fact that in the above proposition we do not get approximation order $m(J)+1$ for the entire Besov class is expected. Indeed, the definition of Besov spaces, say in the univariate case, in terms of divided differences (which is intimately related to the approximation problem we are considering here) involves, for an integer smoothness parameter $k$, a difference operator that is supported on $k+2$ points. Difference operators that involve only $k+1$ points can be used, however, to define the smaller smoothness space $\mathrm{Lip}_k(\mathbb{R})$. One can therefore expect the space $\mathrm{Lip}_{m(J)+1}(\mathbb{R})$ to be the *saturation class* for our problem (in the univariate case).

We show finally that a better approximation rate cannot be obtained for all functions from our Besov space of order $\alpha$. We define a specific $f$ in that space via its Fourier transform, namely

$$\hat{f}(t) = (1 - \chi(t))\|t\|^{-\alpha-n/2}, \qquad t \in \mathbb{R}^n.$$

Here, $\chi$ is the characteristic function of the unit ball. This is in the Besov space because for positive $j$

$$\int_{\Omega_j} \|t\|^{2\alpha}|\hat{f}(t)|^2\, dt = \int_{\Omega_j} \|t\|^{-n}\, dt = \text{const}.$$

Note that $f$ is not contained in the Sobolev space $W_2^\alpha$.

A simple change of variables shows that $\alpha$ is the best order one can achieve:

$$\mathrm{dist}(E^{h\lambda} f, S_h^J(f)) = h^{n/2}\mathrm{dist}(E^\lambda f(h\cdot), S_1^J(f(h\cdot))) \geq h^\alpha \mathrm{dist}(E^\lambda f, S_1^J(f)).$$

Thus, the rate of approximation is $\alpha$, unless $E^\lambda f$ happens to lie in $S_1^J(f)$, i.e. unless $\lambda \in J$.

# 4 Optimal approximation and multivariate polynomial interpolation

Given a finite $J \subset \mathbb{R}^n$, we consider Problem A under the assumption that the underlying function $f$ is 'sufficiently smooth', a notion that we make precise soon. Given $\lambda \in \mathbb{R}^n$, we show that for each $j \in J$ there exists an integer $k_j := k_j(J, \lambda)$, such that, for all sufficiently smooth $f$, the sequence $h^{-k_j}(c_j(\lambda, h))_h$ converges to a finite limit as $h$ tends to zero. Furthermore, under certain assumptions on $J$ and $\lambda$, the limit is shown to be $f$-independent.

Given a finite set $J \subset \mathbb{R}^n$, we assume throughout this section (with the exception of Theorem 4) that our function $f$ of Problem A lies in the Sobolev space $W_2^{M(J)+\varepsilon}(\mathbb{R}^n)$, for some $\varepsilon > 0$. Here, $M(J)$ is the maximal degree of $J$ as defined in Section 2.

EXAMPLE $n = 1$: If $J \subset \mathbb{R}$, then $\Pi_J = \Pi_{m-1}$, with $m$ the cardinality of $J$. We conclude that our smoothness class in one variable is slightly smaller than $W_2^{m-1}(\mathbb{R})$.

As we will see shortly, for a given smooth $f$, the convergence of the optimal coefficient $c_j(\lambda, h)$ depends critically on a certain connection among the three least spaces $\Pi_J$, $\Pi_{(J\backslash j)\cup\lambda}$ and $\Pi_{J\cup\lambda}$. All these spaces are homogeneous, and each one of them is a superspace of $\Pi_{J\backslash j}$. Thus, we can think of each as constructed from $\Pi_{J\backslash j}$ by first appending to that space one (for $J$ and $(J\backslash j)\cup\lambda$) or two (for $J \cup \lambda$) *homogeneous* polynomials, and then taking the span of the polynomial set so obtained. The critical information in this regard is the *degree* of the polynomials appended in such a procedure. This motivates:

**Definition 3** *Let $K$ be a finite subset of $\mathbb{R}^n$ and $k \in \mathbb{R}^n\backslash K$. We denote by*

$$d(K, k)$$

*the degree of any homogeneous polynomial $p$ that satisfies*

$$\Pi_{K\cup k} = \operatorname{span}\{\Pi_K \cup p\}.$$

In general, the value of $d(K, k)$ depends on subtle relations between $K$ and $k$. However, the following estimates (albeit crude ones) are valid:

$$m(K) + 1 \leq d(K, k) \leq M(K) + 1.$$

**Lemma 1** *Let $J$ be a finite subset of $\mathbb{R}^n$, $\lambda \in \mathbb{R}^n$, and $f \in W_2^{M(J)+\varepsilon}(\mathbb{R}^n)$, for some $\varepsilon > 0$. Let $(c_j(\lambda, h))_{j\in J}$ be the optimal coefficients of Problem A. Fix $j \in J$, and let $k_j$ be the $f$-independent integer*

$$k_j = d(J\backslash j, \lambda) - d(J\backslash j, j).$$

*Then the following holds:*

(1) *If $k_j > 0$, $c_j(\lambda, h)$ converges to $0$.*

(2) *If $k_j \leq 0$, there exists a number $c_j(\lambda)$, which is independent of $h$, such that*

$$c_j(\lambda, h) = h^{k_j}(c_j(\lambda) + o(1)),$$

*for $h \to 0$.*

ANALYSIS Already at this point, we can use the statement of the lemma to classify the different cases of Problem A as follows:

(1) The parameter $k_j$ in the lemma is positive. Then, the coefficient $c_j(\lambda, h)$ converges to 0 for every smooth $f$ as $h$ tends to zero.

(2) The parameter $k_j$ in the lemma is negative. Then, the coefficient $c_j(\lambda, h)$ becomes unbounded for a generic smooth $f$ and $h \to 0$.

(3) The parameter $k_j$ is 0, and the principal part $c_j(\lambda)$ is actually independent of $f$. Then, the optimal coefficients converge to *universal limits* $c_j(\lambda)$ that depend on $J$, $\lambda$ and $j$, but not on $f$.

(4) The parameter $k_j$ is 0, but the principal parts depend, in general, on $f$. The optimal coefficients in Problem A converge for every smooth $f$ to the $f$-dependent limit $c_j(\lambda)$.

In case (3) above, a characterisation of the universal coefficients is desired. In any event, the sign of $k_j$ is important. Therefore, it is useful to observe that the space $\Pi_{J\cup\lambda}$ is obtained from $\Pi_{J\backslash j}$ by adding to the latter two homogeneous polynomials: one of degree $d(J\backslash j, j)$ and one of degree $d(J,\lambda)$. Thus, unless $k_j = 0$, it must have the value

$$k_j = d(J,\lambda) - d(J\backslash j, j).$$

While the value of $d(J\backslash j, j)$ may depend on $j \in J$, we have the obvious bound $d(J\backslash j, j) \leq M(J)$. Thus

$$k_j \geq \min\{0, d(J,\lambda)\} - M(J). \tag{4.1}$$

In order to prove Lemma 1, we first state and prove another lemma. All inner products appearing here and elsewhere in this section are standard $L^2$-inner products.

**Lemma 2** *Let $f \in L^2(\mathbb{R}^n)$ and $(g_k)_{k\in J}$ be a finite collection of linearly independent real-analytic functions. Assume that $g_k\widehat{f} \in L^2$, for every $k \in J$. Then the matrix whose entries are*

$$\langle g_j\widehat{f}, g_k\widehat{f}\rangle, \quad (j,k) \in J \times J,$$

*is non-singular.*

**Proof** The non-singularity of the matrix is equivalent to the linear independence of $(g_k\widehat{f})_{k\in J}$; the latter is argued as follows. If

$$\sum_{k\in J} c_k g_k \widehat{f} = 0,$$

then $g\widehat{f} = 0$, with $g = \sum_{k\in J} c_k g_k$. This implies that $g$ must vanish on a non-null subset of $\mathbb{R}^n$ (viz., the support of $\widehat{f}$). Hence $g = 0$ everywhere ($g$ being real analytic). This forces $c_k \equiv 0$ since $(g_k)_k$ are linearly independent by assumption. □

**Proof of Lemma 1** First, we find the coefficient sequence $c(\lambda, h)$ of the best approximation $\sum_{j\in J} c_j(\lambda,h)E^{hj}f$ to $E^{h\lambda}f$ by solving the normal equations. The corresponding Gram matrix is

$$D_h := (\langle E^{hj}f, E^{hk}f\rangle)_{j,k\in J} = (2\pi)^{-n}(\langle e_{hj}\widehat{f}, e_{hk}\widehat{f}\rangle), \tag{4.2}$$

where the rightmost equality follows Parseval's formula. This Gram matrix is not singular, as an application of Lemma 2 with $g_k := e_{hk}$ shows.

Now, we fix $j \in J$, and construct a basis for $\Pi_J$ as follows. First, we choose a basis $B_0 = (b_k)_{k\in(J\backslash j)}$ for $\mathrm{Exp}_{J\backslash j}$ such that (a) the transformation matrix $T_0$ that maps $(e_k)_{k\in(J\backslash j)}$ to $B_0$ is unit lower triangular, and (b) $B_\downarrow := (b_{k\downarrow})_{k\in J\backslash j}$ is a basis for $\Pi_{J\backslash j}$ (such a basis always exists and can be constructed inductively; see (de Boor and Ron, 1992a)). Then we extend this basis to a basis $B$ of $\mathrm{Exp}_J$ by adding an additional function $b$, such that $b_\downarrow$ completes $B_{0\downarrow}$ to a basis $B_\downarrow$ of $\Pi_J$. Let $T$ be the matrix that converts the basis $(e_k)_{k\in J}$ to the basis $B$; by normalising $B$ if necessary, we may assume, as we do, that $\det T = 1$. Then, using (4.2), we obtain that

$$G_h := TD_hT^T = (2\pi)^{-n}(\langle b_i(h\cdot)\widehat{f}, b_k(h\cdot)\widehat{f}\rangle).$$

Due to our smoothness assumption on $f$, it is easy to prove that for small $h$

$$\langle b_i(h\cdot)\widehat{f}, b_k(h\cdot)\widehat{f}\rangle = h^{\deg b_{i\downarrow}+\deg b_{k\downarrow}}(\langle b_{i\downarrow}\widehat{f}, b_{k\downarrow}\widehat{f}\rangle + o(1)).$$

Thus,

$$\det D_h = \det G_h = h^\ell(\det G_\downarrow + o(1)),$$

where $G_\downarrow$ has entries

$$(2\pi)^{-n}\langle b_{i\downarrow}\widehat{f}, b_{k\downarrow}\widehat{f}\rangle, \quad (i,k) \in J \times J,$$

and where

$$\ell := 2\sum_{k\in J\backslash j} \deg b_{k\downarrow} + 2\deg b_\downarrow.$$

The matrix $G_\downarrow$ is non-singular, as follows from Lemma 2, when choosing $(g_k)$ there to be $B_\downarrow$. Thus $h^{-\ell}\det D_h = \det G_\downarrow + o(1)$, $\det G_\downarrow \neq 0$.

We proceed by computing $c_j(\lambda, h)$ via Cramer's rule. The denominator is $\det D_h$. The numerator is the determinant of $D_{h,j}$ which is obtained from $D_h$ when replacing the $j$th column there by

$$(2\pi)^{-n}\langle e_{hi}\widehat{f}, e_{h\lambda}\widehat{f}\rangle,\ i \in J.$$

We apply to $D_{h,j}$ the same row operations as before, i.e., multiply that matrix by $T$ from the left. As to column operations, we need to modify the underlying least space, since the columns now relate to the pointset $J' := (J\backslash j)\cup\lambda$, hence to the least space $\Pi_{J'}$. We construct a basis for $\Pi_{J'}$ by adding to the previous basis $B_0$ of $\mathrm{Exp}_{J\backslash j}$ an additional function $b'$, such that $B' := B_0 \cup b'$ is a basis for $\mathrm{Exp}_{J'}$, while $B'_\downarrow$ is a basis for $\Pi_{J'}$. We let $T_j$ be the matrix that transforms $(e_k)_{k\in J'}$ to $B'$. Then,

$$G_{h,j} := TD_{h,j}T_j^T$$

has entries

$$(2\pi)^{-n}\langle b_i(h\cdot)\widehat{f}, b'_k(h\cdot)\widehat{f}\rangle, \quad (i,k) \in J \times J'.$$

As before, we thus obtain that, with

$$G_{j\downarrow} := (2\pi)^{-n}(\langle b_{i\downarrow}\widehat{f}, b'_{k\downarrow}\widehat{f}\rangle),$$

one has for small $h$

$$h^{-\ell_j} \det G_{h,j} = \det G_{j\downarrow} + o(1),$$

where

$$\ell_j := \deg b_\downarrow + \deg b'_\downarrow + 2 \sum_{k\in J\backslash j} \deg b_{k\downarrow}.$$

Collecting terms, we finally arrive at

$$c_j(\lambda, h) = h^{\deg b'_\downarrow - \deg b_\downarrow} a_j \left(\frac{\det G_{j\downarrow}}{\det G_\downarrow} + o(1)\right),$$

with $a_j := 1/\det T_j$. Since, by the above construction details, $\deg b_\downarrow = d(J\backslash j, j)$ while $\deg b'_\downarrow = d(J\backslash j, \lambda)$, and in view of the discussion in the next paragraph, we obtain the desired result.

In one case, the above argument may fail to go through: this is the case when $\deg b'_\downarrow > M(J)$, a case in which inner products of the form $\langle p\widehat{f}, b'_\downarrow\widehat{f}\rangle$, $\deg p = M(J)$, which necessarily appear in $G_{j\downarrow}$, may not make sense without a stronger smoothness assumption on $f$. However, in such a case $k_j = \deg b'_\downarrow - \deg b_\downarrow \geq \deg b'_\downarrow - M(J) > 0$, and we can modify the previous proof as follows. We extend $B_0$ to a basis $B''$ of $\mathrm{Exp}_{J'}$ such that, with $b''$ the function added to $B_0$, $\deg b''_\downarrow = M(J)$. With this degree reduction, the smoothness assumption on $f$ allows us to invoke the previous argument (with $B''$ replacing $B'$). The new $k_j$ is now $\deg b''_\downarrow - \deg b_\downarrow \geq M(J) - M(J) = 0$. At the same time, $B''_\downarrow$ is now a subset of $\Pi_{J\backslash j}$, hence is dependent, which forces the matrix $G_j$ to be singular. Our previous argument thus yields for that case that $c_j(\lambda, h) = o(1)$, and the proof is now complete. □

We recall the definition of *general position* from Section 2, and remind the reader that this is the generic situation: it is proved in (de Boor and Ron, 1990) that for any integer $\ell$, the sets $J \subset \mathbb{R}^n$ with cardinality $\ell$ that are in general position form an open and dense subset (in $\mathbb{R}^{\ell n}$). Also, note that the degrees $m(J)$ and $M(J)$ of a set $J$ in general position are determined by its cardinality; for example $M(J)$ is the least integer $m$ for which $|J| \leq \dim \Pi_m$. Though general position is the generic case, there are important configurations for $J$ that are not in general position: the most notable case is that of a grid.

**Corollary 1** *Let $J$ be a finite subset of $\mathbb{R}^n$ in general position, $\lambda \in \mathbb{R}^n$, and $f \in W_2^{M(J)+\varepsilon}(\mathbb{R}^n)$, $\varepsilon > 0$. Let $(c_j(\lambda, h))_{j\in J}$ be the optimal coefficients of Problem A. Then, for each $j \in J$, the sequence*

$$h \mapsto c_j(\lambda, h)$$

*converges to a finite limit* $c_j(\lambda)$.

**Proof** In view of Lemma 1, it suffices to show that the integer $k_j$ there is non-negative, for each $j$. Since we have the estimate (4.1), we need only prove that $d(J,\lambda) \geq M(J)$. However, the general position assumption grants us that $\Pi_{M(J)-1} \subset \Pi_J$. Since $d(J,\lambda)$ is the degree of some homogeneous polynomial which is not in $\Pi_J$, this degree is trivially $> M(J)-1$. □

EXAMPLE Let $J$ consist of three points in $\mathbb{R}^2$ which are not collinear. Then $J$ is in general position, hence the corollary applies. The corollary further tells us that, for some $\lambda$ and $j$, the optimal coefficient $c_j(\lambda, h)$ tends to 0. Precisely, the proof of Lemma 2 shows that this happens if $\Pi_{(J\backslash j)\cup\lambda}$ contains a quadratic polynomial (note that $\Pi_J = \Pi_1$). However, the least space of a 3-set contains a quadratic if and only if the three points are collinear. We conclude that $c_j(\lambda, h)$ tends to 0 if $\lambda$ lies on the line that goes through the two other points of $J$. This observation holds for *every* smooth $f$; here, 'smooth' means lying in $W_2^\alpha(\mathbb{R}^2)$, for some $\alpha > 1$.

Suppose now that $J$ is not in general position. First, whether or not $J$ is in general position, it follows from (de Boor and Ron, 1990) that one can always find $j \in J$ such that $d(J\backslash j, j) = M(J)$. On the other hand, for a generic $\lambda$, the value of $d(J\backslash j, \lambda)$ will be the smallest possible, i.e., $m(J\backslash j)+1$. For such $\lambda$ and $j$, with $k_j$ as in Lemma 1,

$$k_j = m(J\backslash j) + 1 - M(J) \leq m(J) + 1 - M(J).$$

This number is non-negative if and *only if* $J$ is in general position. Thus, for $J$ not in general position, the optimal coefficients diverge to $\infty$ for almost all $\lambda$ and $f$. However, one should not mistakenly write off sets $J$ that are not in general position. Such sets can be useful for particular values of $\lambda$. Since we consider $\lambda$ as given, while $J$ is for us to choose, we may adjust $J$ to the value of $\lambda$. Specific examples are given in the sequel.

We now turn our attention to the case when the optimal coefficients of Problem A converge to limits that are $f$-independent. Sufficient conditions for that to happen are given in the next two results. The condition in the first result can be seen to be necessary, too, for the universality of the limit $c_j(\lambda)$. The second result is the main result of this section. Its condition can be shown to be necessary for the existence and universality of *all* coefficient limits.

**Corollary 2** *In Lemma 1, if, for some* $j \in J$, *the least space* $\Pi_{(J\backslash j)\cup\lambda}$ *equals* $\Pi_J$, *then the limit* $c_j(\lambda)$ *of the optimal coefficients* $c_j(\lambda, h)$ *exists and is independent of* $f$. *Moreover, there exists a function* $F \in \mathrm{Exp}_{J\backslash j}$ *such that* $(e_\lambda - (c_j(\lambda)e_j + F))_\downarrow \notin \Pi_J$.

**Proof** Set $J' := (J\backslash j) \cup \lambda$. Since we assume $\Pi_J = \Pi_{J'}$, we may choose in the proof of Lemma 1 the same homogeneous basis $B_\downarrow$ and $B'_\downarrow$ for this space. This shows that $k_j$ in the lemma is 0 and the two matrices $G_\downarrow$ and $G_{j\downarrow}$ are identical. The lemma thus provides the estimate for small $h$

$$c_j(\lambda, h) = a_j + o(1).$$

Thus the optimal coefficient converges to an $f$-independent limit.

We investigate the nature of that limit with the aid of the notations and details introduced in the proof of Lemma 2. We write $b = r_1 e_j + F_1$, and $b' = r_2 e_\lambda + F_2$, $F_1, F_2 \in \mathrm{Exp}_{J\backslash j}$, and assume $b_\downarrow = b'_\downarrow$. If either $F_1$ or $F_2$ is not unique, we choose them to maximise $\deg(b - b')_\downarrow$. We note that $(b - b')_\downarrow \notin \Pi_J$. Indeed, if $g = se_j + F$, $F \in \mathrm{Exp}_{J\backslash j}$, and $g_\downarrow = (b - b')_\downarrow$ then one of the following must happen: (i) $s = 0$. That contradicts the maximal choice of $F_1, F_2$, since $\deg(b - g - b')_\downarrow > \deg(b - b')_\downarrow$. (ii) $s \neq 0$. Then $sb - r_1 g \in \mathrm{Exp}_{J\backslash j}$, and $(sb - r_1 g)_\downarrow = sb_\downarrow = sp$ since $\deg(b - b')_\downarrow > \deg b_\downarrow$. Hence $b_\downarrow \in \Pi_{J\backslash j}$, contrary to our assumptions.

Now, for the matrices $T$ and $T_j$ in Lemma 1, we have that $\det T = r_1$ and $\det T_j = r_2$. Since we assume in the lemma that $\det T = 1$, we have that $r_1 = 1$. Thus, with $a_j = (\det T_j)^{-1} = 1/r_2$, the function $-a_j(b - b')$ is of the form $e_\lambda - (a_j e_j + F_3)$, with $F_3 \in \mathrm{Exp}_{J\backslash j}$, while $(b - b')_\downarrow \notin \Pi_J$. □

**Theorem 3** *Let $J \subset \mathbb{R}^n$ and $\lambda \in \mathbb{R}^n$ be given. If $M(J \cup \lambda) > M(J)$, then*

*(1) The optimal coefficients in Problem A (with the current $J$ and $\lambda$) converge for every $f \in W_2^{M(J)+\varepsilon}$, $\varepsilon > 0$, to an $f$-independent limit $c(\lambda)$.*

*(2) The function $F := \sum_{j \in J} c_j(\lambda) e_j$ is the only function in $\mathrm{Exp}_J$ that satisfies $\deg(e_\lambda - F)_\downarrow > M(J)$, or equivalently, the sequence $j \mapsto c_j(\lambda)$ is the only sequence supported on $J$ for which*

$$p(\lambda) = \sum_{j \in J} c_j(\lambda) p(j), \quad \forall p \in \Pi_{M(J)}.$$

**Proof** Since $M(J \cup \lambda) > M(J)$, we must have $d(J, \lambda) = M(J \cup \lambda) > M(J)$. Now, fix $j \in J$. The space $\Pi_{J \cup \lambda}$ is obtained from $\Pi_{J\backslash j}$ by adding two homogeneous polynomials $p, q$ of degrees $d(J\backslash j, j) \leq M(J)$ and $d(J, \lambda) > M(J)$ respectively. Since these two polynomials were just shown to have different degrees, the space $\Pi_{(J\backslash j)\cup\lambda}$ is obtained from $\Pi_{J\backslash j}$ by appending to the latter either $p$ or $q$ (but not a linear combination of them). We denote by $J_0 \subset J$ those $j \in J$ for which $p$ is the appended polynomial. If $j \in J_0$, then $\Pi_{(J\backslash j)\cup\lambda} = \Pi_J$ (since $\Pi_J$ is also obtained from $\Pi_{J\backslash j}$ by appending either $p$ or $q$; however, $q$ is ruled out because of its higher degree). By Corollary 2, $c_j(\lambda, h)$ converges for every smooth $f$ to an $f$-independent limit. In the opposing case when the appended polynomial is $q$, then $d(J\backslash j, \lambda) = d(J, \lambda) = \deg q > M(J)$, and hence $k_j$ of Lemma 1 is *positive*. In this case $c_j(\lambda, h) \to 0$.

We now prove the characterisation of the limit coefficients. Let $q$ be as in the previous paragraph. Let $F_0 \in \mathrm{Exp}_{J\cup\lambda}$ be such that $F_{0\downarrow} = q$. $F_0$ is unique: if $\mathcal{F}_{0\downarrow} = q$ as well, and $F_0 - \mathcal{F}_0 \neq 0$, then $(F_0 - \mathcal{F}_0)_\downarrow$ is a polynomial in $\Pi_{J\cup\lambda}$ of degree $> \deg q$, which is impossible. However, we have already proved that, for $j \in J\backslash J_0$, $q \in \Pi_{(J\backslash j)\cup\lambda}$; hence the uniqueness of $F_0$ implies that $F_0 \in \mathrm{Exp}_{(J\backslash j)\cup\lambda}$. Consequently, $F_0 \in \mathrm{Exp}_{J_0\cup\lambda}$.

The coefficient of $e_\lambda$ in $F_0$ is not zero, since otherwise $F_{0\downarrow} \in \Pi_J$, and $\Pi_J$ cannot contain a polynomial whose degree exceeds its own maximal degree. Normalising if necessary, we may assume that $F_0 = e_\lambda - F$, for some $F \in \mathrm{Exp}_J$. Note that $F_0$ as above is the unique function in $\mathrm{Exp}_J$ for which $F_{0\downarrow} \notin \Pi_J$. Indeed, if $\mathcal{F}_0 = e_\lambda - \mathcal{F}$, for some $\mathcal{F} \in \mathrm{Exp}_J\backslash\{F\}$, then $\mathcal{F}_0 - F_0 = F - \mathcal{F} \in \mathrm{Exp}_J\backslash 0$, and hence $\deg(\mathcal{F}_0 - F_0)_\downarrow \leq M(J)$. This implies that $\deg \mathcal{F}_{0\downarrow} \leq M(J)$. Since $\deg F_{0\downarrow} > M(J)$, we have found in $\Pi_{J\cup\lambda}\backslash\Pi_J$ two homogeneous polynomials of different degrees, and that contradicts the fact that $\dim \Pi_{J\cup\lambda} = \dim \Pi_J + 1$.

We write $F = \sum_{j\in J} d_j e_j$, and will show that $c_j(\lambda) = d_j$ for every $j$. Since $F$ lies in $\mathrm{Exp}_{J_0}$, we have $d_j = 0$, for every $j \in J\backslash J_0$. This agrees with our previous observation that $c_j(\lambda) = 0$ for each such $j$. In the opposite case, $j \in J_0$. Corollary 2 and the uniqueness of $F$ (proved in the previous paragraph) show $d_j = c_j(\lambda)$. □

**Proof of Theorem 1** With $m := m(J) = M(J)$, we know that $\Pi_J = \Pi_m$. Since $\Pi_{J\cup\lambda}$ is a proper superspace of $\Pi_J$, it must contain polynomials of degree $> m$. Theorem 3 applies to yield the first part of Theorem 1.

From the corollary we also know that, with $c(\lambda)$ the limit coefficients, $\deg(e_\lambda - \sum_{j\in J} c_j(\lambda)e_j)_\downarrow > m$, and hence that $|E_{c(\lambda)}(t)| = O(\|t\|^{m+1})$ near the origin. Part (2) follows by invoking Theorem 2. □

One may try to adapt $J$ to the given $\lambda$ and $f$ in a way that admits the use of a small $J$ on the one hand, as well as the relaxation of the smoothness assumption on $f$ on the other hand. In the statement below, the symbol $D_\theta$ stands for the directional derivative in the $\theta$-direction.

**Theorem 4** *Given $\theta \in \mathbb{R}^n$, $m > 0$ and $f_0 \in L^2(\mathbb{R}^n)$, assume that $D_\theta^m f_0 \in L^2(\mathbb{R}^n)$. Given any $\lambda \in \mathbb{R}^n$, let $J$ be a subset of $\mathbb{R}^n$ of cardinality $m$ such that $J \cup \{\lambda\}$ lies on a line in the $\theta$ direction. Then, there exists a sequence $c(\lambda) \in \mathbb{C}^J$, that depends only on $\lambda$ and $J$, with the following properties:*

*(1) The optimal coefficients of Problem A w.r.t. $f_0$ converge to $c(\lambda)$.*

*(2) For every $\alpha \leq m$, and every $f \in L^2(\mathbb{R}^n)$, the $L^2(\mathbb{R}^n)$ error in*

$$E^{h\lambda} f \approx \sum_{j\in J} c_j(\lambda) E^{hj} f$$

*is $O(h^\alpha)$, with $\alpha$ the maximal integer for which $D_\theta^\alpha f \in L^2(\mathbb{R}^n)$.*

We note that (de Boor and Ron, 1990), if a set $J$ lies entirely on a line directed in the $\theta$ direction, then $\Pi_J = \text{span}\{t \mapsto (\theta \cdot t)^\ell \mid 0 \le \ell \le |J|-1\}$. In particular, for a collinear set $J$, $M(J) = |J| - 1$.

**Proof of Theorem 4** Since we assume that $J \cup \lambda$ is collinear, so is $J$, and hence, by the remark preceding this proof, $M(J \cup \lambda) = |J| > |J| - 1 = M(J)$. Part (1) of the theorem then follows from Theorem 3, as soon as we verify that the weaker smoothness assumption imposed on $f$ is suitable.

The fact that $f$ in Corollary 1 is required to be smooth is due to the estimation in its proof of expressions of the form $\int (b - b_\downarrow)|\widehat{f}|^2$, for certain $b \in \text{Exp}_{J\cup\lambda}$. We used there the estimate $|(b - b_\downarrow)(t)| \le \text{const}\|t\|^{\deg b_\downarrow + 1}$. However, in the present case, $|b - b_\downarrow|$ is a function of the variable $t \mapsto \theta \cdot t$. Hence we can bound $|(b - b_\downarrow)(t)| \le \text{const}|\theta \cdot t|^{\deg b_\downarrow + 1}$, and that allows us to relax our smoothness condition on $f$.

In order to prove (2), we invoke the characterisation of the principal parts of Theorem 3: the characterisation says that, with $c(\lambda)$ being the universal limits of the optimal coefficients, we have $\deg E_{c(\lambda)\downarrow} = M(J) + 1 = |J|$. This implies $|E_{c(\lambda)}(t)| = O((\theta \cdot t)^{|J|})$. By an argument analogous to that employed in the proof of Theorem 2, we obtain that the approximation scheme has the properties asserted in (2). □

EXAMPLE: RECTANGULAR GRIDS If $J \cup \lambda$ is the cartesian product of univariate sets $(J_1, \ldots, J_n)$ of cardinalities $(k_1, \ldots, k_n)$, then (de Boor and Ron, 1992a) $\Pi_{J\cup\lambda}$ is spanned by the monomials

$$t \mapsto t^\alpha, \quad \alpha \le k := (k_1 - 1, \ldots, k_n - 1).$$

The same reference shows that, upon deleting the monomial $t \mapsto t^k$ from the above basis, we get a basis for $\Pi_{J'}$ with $J'$ obtained from $J \cup \lambda$ by the deletion of *any* single point in $J \cup \lambda$. In particular, $M(J \cup \lambda) > M(J)$. Thus, Theorem 3 applies to show that the optimal coefficients converge here to universal limits.

We close this section with an example showing that the requirement $M(J \cup \lambda) > M(J)$ (cf. Theorem 3) is not *necessary* for the universality of the limits of *some* of the optimal coefficients.

EXAMPLE Assume that $J_i$ is a finite subset of the $x_i$ axis, $i = 1, 2$, of $\mathbb{R}^2$, $J = J_1 \cup J_2$. Then $\Pi_J$ is spanned by pure monomials (i.e., by functions of the form $t \mapsto t_i^m$). For $\lambda$ on the $x_1$-axis, $\Pi_{J\cup\lambda}$ is obtained from $\Pi_J$ by appending $t \mapsto t_1^{|J_1|}$ to $\Pi_J$, and thus $d(J, \lambda) = |J_1|$. However, if $|J_2| > |J_1|$, then $M(J) = |J_2| - 1 \ge |J_1|$, and hence $M(J \cup \lambda) = M(J)$. At the same time, for each $j \in J_1$, we have that $\Pi_{(J\setminus j)\cup\lambda} = \Pi_J$, hence that $c_j(\lambda, h)$ converges, for such $j$ and for all smooth $f$, to an $f$-independent limit. In summary, in this example, despite the fact that the condition required in Theorem 3 fails to hold, Corollary 2 still applies to show that some of the coefficients converge to universal limits.

# 5 Optimal approximation and radial basis functions

In the previous section, a fairly thorough analysis of the convergence of the optimal coefficients in Problem A is provided when $f$ is 'sufficiently smooth'. In many cases, we can easily circumvent the smoothness assumption on $f$ by simply reducing the cardinality of $J$: the smoothness assumption is relaxed as we remove points from $J$. However, this clearly entails that, in such case, the optimal coefficients may converge, but at the expense of an increase in the error of best approximation. After all, the spaces $(S_h^J())_h$ ought to approximate worse as $J$ is reduced.

The purpose of this section is to give sufficient conditions for the existence of the principal parts of the optimal coefficient sequences for functions that are not covered by the results of the previous section.

To begin, we state a set of conditions on $f$, expressed in terms of its Fourier transform. They will apply to the next theorem, and they are related to, but more general than (1.4). To that end, we require that $f : \mathbb{R}^n \to \mathbb{R}$ is absolutely integrable and square-integrable. Its Fourier transform $\hat{f}$ is therefore continuous. We require it to be *slowly varying*, i.e. there is a positive function $G$ such that

$$\lim_{\|y\|\to\infty} \frac{\hat{f}(t\|y\|)}{\hat{f}(y)} = G(t), \qquad t \text{ a.e. in } \mathbb{R}^n. \tag{5.1}$$

Here 'a.e.' means almost everywhere, i.e. everywhere except perhaps on a set of measure zero.

In order that the convergence (5.1) is controlled, we require that there is a constant $K$ such that

$$\left|\frac{\hat{f}(t\|y\|)}{\hat{f}(y)}\right| \leq KG(t) \tag{5.2}$$

for almost all $t \in B :=$ the unit ball. To explain the relevance of conditions (5.1)–(5.2), note that in the univariate case, they are a natural generalisation of (1.4).

We finally have a condition that fixes certain properties of $G$'s behaviour at the origin and for large argument. Its purpose is to render the limiting minimisation problem well-defined. We state this condition by viewing $G$ as the Fourier transform of a certain generalised function, and it will be convenient to deal with the square of $G$ instead of $G$ itself. Precisely, we require that the square of $G$ is the distributional Fourier transform of a function $\phi$ in $\mathcal{P}$, a class with the following properties.

Every $\phi \in \mathcal{P}$ must have a distributional Fourier transform which agrees with a positive function $\hat{\phi} \in C(\mathbb{R}^n \setminus 0) \cap L^1(\mathbb{R}^n \setminus B)$. Further, we require that there exists a sequence $c = (c_j)_{j\in J}$, such that for all $\lambda \in \mathbb{R}^n$

$$\int_{\mathbb{R}^n} |E_c(t)|^2 \hat{\phi}(t)\, dt < \infty. \tag{5.3}$$

$E_c$ still has the same meaning as in the previous sections.

EXAMPLE An example where all these assumptions are satisfied is when $\hat{f}$ is a negative power of the function $(1 + \|\cdot\|)$ with a suitably large negative exponent $-\alpha - n/2$ to ensure $G^2 = \|\cdot\|^{-2\alpha-n} = \hat{\phi} \in L^1(\mathbb{R}^n \setminus B)$. We note, however, that (5.1) is really an asymptotic condition and does not require $\hat{f}$ to be of that form. If, as in this example, $\hat{\phi}(t) = \|t\|^{-2\alpha-n}$, then the aforementioned $\phi(t)$ is a constant multiple of $\|t\|^{2\alpha}$ so long as $\alpha$ is *not* an integer, whereas in the opposing case $\phi(t)$ is a constant multiple of $\|t\|^{2\alpha}\log\|t\|$. Both of these $\phi$ are in the class of radial basis functions frequently considered in the literature (Buhmann, 1993, Dyn, 1989, Micchelli, 1986). In order to satisfy (5.3) for this example, we have to choose $c$ such that

$$\sum_{j\in J} c_j j^\gamma = \lambda^\gamma, \qquad |\gamma| \le \alpha. \tag{5.4}$$

Then the condition (5.3) is true. It is always possible to achieve the above requirement by taking a large enough $J$.

The relevance of radial basis functions in general in this section will become clear in the following theorem, where we demonstrate that under our conditions, the limiting coefficients are Lagrange functions of certain radial basis function interpolation problems. We point out that $J$ is still a set of scattered points in $\mathbb{R}^n$. It is a salient assumption that the only polynomial from the kernel $\mathcal{K}$ of the semi inner product associated with $\hat{\phi}$, viz.,

$$\langle f, g\rangle = \int_{\mathbb{R}^n} \hat{\phi}^{-1} \hat{f}\bar{\hat{g}},$$

that vanishes identically on $J$, is the zero polynomial. The above inner product is well-defined for all $f, g \in \mathcal{H}$, where

$$\mathcal{H} := \left\{ f \in \mathcal{S}' \mid \int_{\mathbb{R}^n} \hat{\phi}^{-1} |\hat{f}|^2 < \infty \right\}.$$

Here, $\mathcal{S}'$ is the space of distributions dual to the Schwartz space of rapidly decreasing smooth test functions on $\mathbb{R}^n$, see, e.g., (Jones, 1982).

The above assumptions on the points we make from now on. We also find it convenient for the statement of the next theorem to define the space

$$\mathcal{J} := \left\{ d = (d_j)_{j\in J} \in \mathbb{R}^J \mid \sum_{j\in J} d_j q(j) = 0\, \forall\, q \in \mathcal{K} \right\}.$$

Our last result now is as follows.

**Theorem 5** *Let $f$ be such that the above conditions* (5.1)–(5.2) *hold for $G = \sqrt{\hat{\phi}}$, $\phi \in \mathcal{P}$. Then, with $(c_j(\lambda, h))_{j\in J}$ the optimal coefficients of Problem A, the following limits exist:*

$$\lim_{h\to 0} c_j(\lambda, h) = c_j(\lambda), \qquad j \in J, \qquad \lambda \in \mathbb{R}^n,$$

*and $(c_j(\lambda))_{j\in J}$ are the coefficients that minimise*

$$\min \int_{\mathbb{R}^n} |E_c(t)|^2 \hat{\phi}(t)\, dt \tag{5.5}$$

*over all coefficients satisfying* (5.3)*. They are also the unique Lagrange functions in $\lambda$ of the form*

$$c_j(\lambda) = \sum_{k\in J} d_{jk}\phi(\lambda - k) + p(\lambda), \qquad \lambda \in \mathbb{R}^n,\ j \in J,$$

*that provide the interpolation conditions*

$$c_j(k) = \delta_{jk}, \qquad j \in J,\ k \in J,$$

*where $p$ is a polynomial from $\mathcal{K}$ and $(d_{jk})_{k\in J} \in \mathcal{J}$.*

**Proof** First note that, due to Parseval's identity, the norm (1.3) that we need to minimise equals the square root of

$$\frac{1}{(2\pi)^n} \int_{\mathbb{R}^n} |E_c(th)|^2 |\hat{f}(t)|^2\, dt, \tag{5.6}$$

and hence we are entitled to minimise, in lieu of (1.3), the above expression. To this end, we multiply (5.6) by

$$(2\pi h)^n |\hat{f}(h^{-1}\mathbf{1}/\sqrt{n})|^{-2},$$

where $\mathbf{1} = (1, 1, \ldots, 1) \in \mathbb{R}^n$, and scale the argument in the integral to obtain

$$|\hat{f}(h^{-1}\mathbf{1}/\sqrt{n})|^{-2} \int_{\mathbb{R}^n} |E_c(t)|^2 |\hat{f}(h^{-1}t)|^2\, dt \tag{5.7}$$

instead of (5.6); thus, we can minimise (5.7), instead of (1.3). Now, let $\mathbf{c}$ be the minimising sequence of (5.5), and let $c(\lambda, h)$ be the minimising sequence of (5.7), the latter being the same as the minimising sequence of (1.3) (the uniqueness of these minimising sequences is guaranteed by Lemma 2). Then, with $c_0$ any finite accumulation point of $(c(\lambda, h))_h$, and with $C_k := c(\lambda, h_k)$ a subsequence that converges to $c_0$, we get from (5.1) that

$$|\hat{f}(h_k^{-1}\mathbf{1}/\sqrt{n})|^{-2} |E_{C_k}(t)|^2 |\hat{f}(h_k^{-1}t)|^2$$

converges pointwise to $|E_{c_0}(t)|^2\widehat{\phi}(t)$. Therefore, by Fatou's Lemma,

$$\int_{\mathbb{R}^n} |E_{c_0}(t)|^2\widehat{\phi}(t)\,dt \leq \liminf_{k\to\infty} |\hat{f}(h_k^{-1}\mathbf{1}/\sqrt{n})|^{-2} \int_{\mathbb{R}^n} |E_{c_k}(t)|^2 |\hat{f}(h_k^{-1}t)|^2\,dt. \quad (5.8)$$

On the other hand, for any $h > 0$, since we are assuming that $\mathbf{c}$ satisfies (5.3), the following bound is valid:

$$|\hat{f}(h^{-1}\mathbf{1}/\sqrt{n})|^{-2}|E_{\mathbf{c}}(t)|^2|\hat{f}(h^{-1}t)|^2 \leq \begin{cases} |E_{\mathbf{c}}(t)|^2\hat{\phi}(t) + \text{const}, & \text{if } \|t\| \leq 1, \\ \hat{\phi}(t)\cdot\text{const}, & \text{if } \|t\| \geq 1. \end{cases}$$

By the dominated convergence theorem, we thus get that

$$\lim_{h\to 0} |\hat{f}(h^{-1}\mathbf{1}/\sqrt{n})|^{-2} \int_{\mathbb{R}^n} |E_{\mathbf{c}}(t)|^2 |\hat{f}(h^{-1}t)|^2\,dt = \int_{\mathbb{R}^n} |E_{\mathbf{c}}(t)|^2\widehat{\phi}(t)\,dt.$$

Combining this with (5.8), we conclude that, in view of the optimality of $c(\lambda, h)$,

$$\int_{\mathbb{R}^n} |E_{c_0}(t)|^2\widehat{\phi}(t)\,dt \leq \int_{\mathbb{R}^n} |E_{\mathbf{c}}(t)|^2\widehat{\phi}(t)\,dt.$$

This, in turn, implies, since $\mathbf{c}$ is the unique solution of (5.5), that $c_0 = \mathbf{c}$; in other words, the sequence $(c(\lambda, h))_h$ has $\mathbf{c}$ as its unique finite accumulation point. This leaves us to show just the boundedness of the limiting coefficients to settle the lemma.

Indeed, in the following fashion we can bound the optimal coefficients independently of $h$. Let, for the time being, $g : \mathbb{R}^n \to \mathbb{R}$ be *any* square-integrable function that satisfies the following conditions for a fixed $k \in J$:

$$g(\lambda) = 0, \quad g(j) = \delta_{jk}, \qquad j \in J,$$

where we assume $\lambda \neq j$ for all $j \in J$ (otherwise the solution of the minimisation problem would be obvious). Such a function exists of course. Then, by Cauchy–Schwarz,

$$\begin{aligned} |c_k(\lambda, h)|^2 &= \Big|g(\lambda) - \sum_{j\in J} c_j(\lambda, h)g(j)\Big|^2 \\ &\leq \frac{1}{(2\pi)^{2n}}\left[\int_{\mathbb{R}^n} |\hat{g}(t)||E_{c(\lambda,h)}(t)|\,dt\right]^2 \\ &\leq \frac{1}{(2\pi)^{2n}}\int_{\mathbb{R}^n} \frac{|\hat{g}(t)|^2}{|\hat{f}(h^{-1}t)|^2}\,dt \int_{\mathbb{R}^n} |E_{c(\lambda,h)}(t)|^2|\hat{f}(h^{-1}t)|^2\,dt \\ &= \frac{1}{(2\pi)^{2n}}\int_{\mathbb{R}^n} \frac{|\hat{g}(t)|^2}{|\hat{f}(h^{-1}\mathbf{1}/\sqrt{n})|^{-2}|\hat{f}(h^{-1}t)|^2}\,dt \,\times \\ &\quad |\hat{f}(h^{-1}\mathbf{1}/\sqrt{n})|^{-2}\int_{\mathbb{R}^n} |E_{c(\lambda,h)}(t)|^2|\hat{f}(h^{-1}t)|^2 dt. \end{aligned}$$

We take $g(t) = p(t)\psi(t-k)$, where $p$ is a polynomial of suitable degree, $p(\lambda) = 0$, $p(j) = \delta_{jk}$, $\psi$ is entire and quickly decaying along $\mathbb{R}^n$, so that $\hat{\psi}$ is sufficiently smooth, with $\hat{\psi}$'s support in $B$, and $\psi(0) = 1$. Such a polynomial exists; we may form it, for instance, as a suitably scaled product of terms of the form $\|t-j\|^2$, $\|t-\lambda\|^2$. We get, using the properties of $g$ and $\hat{f}$,

$$\limsup_{h\to 0} |c_k(\lambda,h)|^2 \le \text{const} \cdot \int_{\|t\|<1} \hat{\phi}(t)^{-1}\,dt \times$$
$$\limsup_{h\to 0} |\hat{f}(h^{-1}\mathbf{1}/\sqrt{n})|^{-2} \int_{\mathbb{R}^n} |E_c(t)|^2 |\hat{f}(h^{-1}t)|^2\,dt$$

which is uniformly finite.

Finally, we explain the representation of the principal parts as Lagrange functions. Indeed, it is clear from (1.3) that the principal parts $c_j(\lambda)$ as $h \to 0$ of the optimal coefficients $c_j(\lambda,h)$ are fundamental functions with respect to $\lambda$ on the set $J$, i.e. they must yield the interpolation conditions

$$c_j(k) = \delta_{jk}, \qquad j \in J,\ k \in J.$$

Further, it follows from the fact that the principal parts solve (5.5) and from $\phi \in \mathcal{P}$ that they are of the form

$$c_j(\lambda) = \sum_{k\in J} d_{jk}\phi(\lambda - k) + p(\lambda), \qquad \lambda \in \mathbb{R}^n,$$

where $p$ is an element of the kernel $\mathcal{K}$ of the aforementioned semi inner product $\langle\cdot,\cdot\rangle$ associated with $\hat{\phi}$. Moreover, the $d_{jk}$ have to satisfy the side conditions mentioned in the statement of the theorem. This fact is a consequence of standard Hilbert space theory for positive definite kernels on subspaces of $\mathbb{R}^J$, see (Dyn, 1989) and (Schaback, 1993), for good summaries of this topic. The subspace here is $\mathcal{J}$. □

We observe immediately that the coefficients of this theorem give a scheme to which Theorem 2 can be applied, for instance in the example given at the beginning of this section. In that case, i.e. when $\hat{\phi}(t) = \|t\|^{-2\alpha-n}$, $\mathcal{K}$ contains $\Pi_{<\alpha+n/2}$ which means that all such polynomials are reproduced by the Lagrange interpolation of Theorem 5 (by uniqueness of interpolation). Therefore, (5.4) holds even for all $|\gamma| < \alpha + n/2$ using the $c_j(\lambda,h)$. Theorem 2 is thus applicable for $m \le \lceil \alpha + n/2 \rceil$, the $f$ given in the example being in $B^\alpha_\infty(L^2(\mathbb{R}^n))$.

# References

[1] de Boor, C. and A. Ron (1990) "On multivariate polynomial interpolation", *Constr. Approx.* **6**, 287–302.

[2] de Boor, C. and A. Ron (1992a) "Computational aspects of polynomial interpolation in several variables", *Maths of Comput.* **58**(198), 705–727.

[3] de Boor, C. and A. Ron (1992b) "The least solution of the multivariate polynomial interpolation", *Math. Z.* **210**, 347–378.

[4] Buhmann, M.D. (1993) "New developments in the theory of radial basis function interpolation", in *Multivariate Approximation: From CAGD to Wavelets* (K. Jetter and F.I. Utreras, eds), World Scientific (Singapore), 35–75.

[5] Buhmann, M.D. and C.A. Micchelli (1992) "On radial basis approximation on periodic grids", *Math. Proc. Cambridge Phil. Soc.* **112**, 317–334.

[6] Dyn, N. (1989) "Interpolation and approximation by radial and related functions", in *Approximation Theory VI* (C.K. Chui, L.L. Schumaker, and J.D. Ward, eds), Academic Press (New York), 211–234.

[7] DeVore, R.A. and G.G. Lorentz (1993) *Constructive Approximation,* Springer-Verlag.

[8] Iserles, A. and G. Strang (1983) "The optimal accuracy of difference schemes", *Trans. Amer. Math. Soc.* **277**, 779–803.

[9] Jones, D.S. (1982) *The Theory of Generalised Functions,* Cambridge University Press (Cambridge, UK).

[10] Micchelli, C.A. (1986) "Interpolation of scattered data: distance matrices and conditionally positive definite functions", *Constr. Approx.* **1**, 11–22.

[11] Micchelli, C.A. and W.L. Miranker (1973a) "Optimal difference schemes for linear initial value problems", *SIAM J. Numer. Anal.* **10**, 983–1009.

[12] Micchelli, C.A. and W.L. Miranker (1973b) "Asymptotically optimal approximation in fractional Sobolev spaces and the numerical solution of differential equations", *Numer. Math.* **22**, 75–87.

[13] Schaback, R. (1993) "Comparison of radial basis function interpolants", in *Multivariate Approximation: From CAGD to Wavelets* (K. Jetter and F.I. Utreras, eds), World Scientific (Singapore), 293–305.

# On the Convergence of Derivative-Free Methods for Unconstrained Optimization

*Andrew Conn*

IBM T.J. Watson Research Center, P.O.Box 218, Yorktown Heights, NY 10598, USA

*Katya Scheinberg*

Industrial Engineering and Operations Research Department, Columbia University, New York, NY 10027-6699, USA

*Philippe Toint*

Department of Mathematics, Facultés Universitaires ND de la Paix, 61, rue de Bruxelles, B-5000 Namur, Belgium

**Abstract**

The purpose of this paper is to examine a broad class of derivative-free trust-region methods for unconstrained optimization inspired by the proposals of [21] and to derive a general framework in which reasonable global convergence results can be obtained. The developments make extensive use of an interpolation error bound derived in [23] in the context of multivariate polynomial interpolation.

## 1 Introduction

The problem that we consider in this paper is that of minimizing a nonlinear smooth objective function of several variables when the derivatives of the objective function are unavailable. More formally, we consider the problem

$$\min_{x \in \mathbb{R}^n} f(x), \tag{1.1}$$

where we assume that $f$ is smooth and that $\nabla f(x)$ (and, a fortiori, $\nabla^2 f(x)$) cannot be computed for any $x$.

The main motivation for examining algorithmic solutions to this problem is the high demand from practitioners for such tools. In the cases presented to the authors, $f(x)$ is typically *very expensive* to compute, and its derivatives are not available either because $f(x)$ results from some physical, chemical or econometrical measure, or, more commonly, because it is the result of a possibly very large computer simulation, for which the source code is effectively unavailable. The occurrence of problems of this nature appears to be surprisingly frequent in the industrial world.

When users are faced with such problems, there are several strategies that can be considered. The first, and maybe simplest, is to apply existing 'direct search' optimization methods, like the well-known and widely used simplex reflection algorithm of [13] or its modern variants, or the Parallel Direct Search algorithm of [8] and [28]. This first approach has the merit of requiring little additional effort from the user, but may require substantial computing resources: the inherent smoothness of the objective function is not very well exploited, and, as a result, the number of function evaluations is sometimes very large, a major drawback when these evaluations are expensive.

The second and more sophisticated approach is to turn to automatic differentiation tools (see [9] or [10], for instance). However, such tools are unfortunately not applicable in the two typical cases mentioned above since they require the function that is to be differentiated be the result of a callable program that cannot be treated as a black box.

A third possibility is to resort to finite difference approximation of the derivatives (gradients and possibly Hessian matrices). A good introduction to these techniques is provided by the book [7], for instance. In general, given the cost of evaluating the objective function, evaluating its Hessian by finite differences is much too expensive; one can use quasi-Newton Hessian approximation techniques instead. In conjunction with the use of finite differences for computing gradients, this type of method has proved to be useful and sometimes surprisingly efficient.

We will however focus this paper on a fourth possible approach, which is based on the idea of modelling the objective function directly, instead of modelling its derivatives. This idea seems particularly attractive in that one can replace an expensive function evaluation by a much cheaper surrogate model and, especially for very complex problems, make considerable progress in obtaining improved solutions at moderate cost. Such modelling has been promoted by several contributions of Powell. The following interesting argument is for instance proposed in [18]:

> However I believe that eventually the better methods will not use derivative approximations. ... One reason for my belief comes from the relatively easy problem of solving a single nonlinear equation. Here I think it is fair to compare the secant method with the Newton–Raphson method. The secant method requires one function evaluation per iteration and has a convergence rate of 1.618 per function value

> ... while the Newton–Raphson method requires one function and one derivative value per iteration and has quadratic convergence. Therefore, if an extra function value is used to estimate the derivative in the Newton–Raphson iteration, the mean rate of convergence per function value is about 1.414, which is not as good as the secant method.

We will call *derivative-free optimization* the techniques which do not attempt to directly compute approximations to the unavailable derivative information, but rather derive improvements from a model of the objective function itself without this derivative information. Our purpose in this paper is to explore the theoretical aspects of these methods: we will define a broad class of derivative-free trust-region methods and consider their convergence theory.

The next section will be devoted to Powell's contributions to derivative-free optimization, because of their crucial nature and the influence they have had on the rest of our presentation. Section 3 will be concerned with the description of the algorithmic framework we wish to study, while convergence of this framework will be examined in Section 4. Some conclusions and perspectives will finally be outlined in Section 5.

# 2 Powell's contributions to derivative-free optimization

Since the 1960s, Powell's views on derivative-free optimization have had a substantial influence on the development of this area, both because of their varied and their fundamental nature.

## 2.1 Conjugate directions

The first major contribution of Powell in this context appeared in [14], where he described a method for solving the nonlinear unconstrained minimization problem based on the use of conjugate directions. The main idea is that the minimum of a positive-definite quadratic form can be found by performing at most $n$ successive line searches along mutually conjugate directions, where $n$ is the number of variables. The same procedure may of course be applied to nonquadratic functions, adding a new composite direction at the end of each cycle of $n$ line searches. Of course, finite termination is no longer expected in this case.

This algorithm has enjoyed a lot of interest amongst both numerical analysts and practitioners. The properties of the method were analyzed in [3], [4], [27] and [26]. On the more practical side, Powell made his method available in the Harwell Subroutine Library under the name of VA04. The original routine is no longer distributed, and has been replaced by VA24, also written by Powell (see [11]). Brent also proposed a variant of the algorithm and gave an ALGOL W code named PRAXIS. This code was subsequently translated into Fortran by R. Taylor, S. Pinski and J. Chandler, and has since been widely used in practice[1].

Methods of conjugate directions have proved to be reasonably reliable, but suffer from two main disadvantages: the need to maintain good linear independence of the search directions and the relative difficulty of determining near-conjugate directions when the Hessian of the function is ill-conditioned.

Recognizing these difficulties, Powell [18] suggested using orthogonal transformations of sets of conjugate directions, a technique leading to the idea of B-conjugate methods, as described in [19]. The main idea of this proposal was to approximate the matrix of second derivatives itself, by modifying an initial estimate to ensure that it satisfies properties which would be satisfied if the objective function were quadratic. He also suggested the use of variational criteria, such as those used to derive quasi-Newton updates, in order that good information from an approximation can be inherited by its successors at subsequent iterations.

We wish to emphasize here that this approach is already aimed at building approximate (quadratic) models for the objective function, an idea which is crucial in our version of derivative-free optimization methods.

## 2.2 Methods for nonlinear equations

Powell also provided significant input in the related field concerned with the numerical solution of sets of nonlinear equations and nonlinear least-squares problems. The method described in [15] again uses uses the idea of searching along a set of linearly independent directions, but estimates the derivatives of the residuals along these directions by finite differences, thereby avoiding the use of analytical expressions for derivatives. A search direction is then computed using the Gauss–Newton equation, and is used to replace one search direction of the current set. The method exploits the fact that successive Gauss–Newton matrices only differ by a low rank term.

Five years later, [16] describes a method where the Broyden quasi-Newton formula is used to approximate the Jacobian matrices associated with such problems, promoting at the same time the idea of a modelling the nonlinear function of interest within a trust region, an idea that we will pursue in the rest of this paper. The implementation of this algorithm for the Harwell Subroutine Library under the name of NS01 is detailed in [17], and was for a long time a favourite amongst practitioners and numerical analysts alike.

[1] PRAXIS is distributed in the public domain by J. Chandler, Computer Science Department, Oklahoma State University, Stillwater, Oklahoma 70078, USA (jpc@a.cs.okstate.edu). There is also an interface with the CUTE testing environment of [1].

## 2.3 Methods based on multivariate interpolation

The third important contribution by Powell is the main source of our inspiration for the rest of this paper. [20] proposed a method for constrained optimization, where the objective function and constraints are approximated by linear multivariate interpolation. At the same time, the associated Fortran code, named COBYLA, was (and still is) distributed by Powell to interested parties[2]. A little later, [21] described an algorithm for unconstrained optimization using a multivariate quadratic interpolation model of the objective function, an approach revisited by [22].

The main idea of Powell's proposal was to use the available objective function values $f(x_i)$ (or in the constrained case, an $\ell_\infty$ penalty function). For the purposes of this paper we will confine our attention to the unconstrained results where an interpolation model for $f$ is used. This model is assumed to be valid in a neighbourhood of the current iterate: this neighbourhood is described as a trust region (a hypersphere centred at $x_i$ and of radius $\Delta$). This model is then minimized within the trust region, hopefully yielding a point with a low function value. As the algorithm proceeds and more objective function values become available, the points defining the interpolation model (the interpolation set) are updated in a way that preserves their geometrical properties and the trust-region radius is also adapted. Of course, this is only a very crude description of Powell's idea, but we will return to a more formal framework in the next section.

A variant of Powell's quadratic interpolation scheme is discussed in [6], where encouraging numerical results are presented. Similar computational results were also mentioned in [22].

# 3 A class of trust region algorithms for derivative-free optimization

We now turn to a more precise description of the problem and of the algorithmic framework we propose to analyze. We consider problem (1.1) and assume that $f(x)$ can be evaluated (at a supposedly high cost) for every $x \in \mathbb{R}^n$. As in [21], [6] and [22], we propose to use an iterative algorithm of the trust-region type, in which the objective function is modelled, at iteration $k$, by a quadratic of the form

$$m_k(x_k + s) = f(x_k) + \langle g_k, s \rangle + \tfrac{1}{2}\langle s, H_k s \rangle. \tag{3.1}$$

In this definition, we have used the notation $\langle \cdot, \cdot \rangle$ to denote the usual Euclidean inner product on $\mathbb{R}^n$, and we will subsequently use the symbol $\| \cdot \|$ for the associated Euclidean norm. We have also used $x_k$ to denote the $k$-th

[2] An interface with the CUTE environment of [1] is also provided for COBYLA.

iterate, while $g_k$ is a vector of $\mathbb{R}^n$ and $H_k$ a symmetric $n \times n$ matrix to be determined. The vector $s$ represents a step from the iterate $x_k$, which, according to the trust-region methodology, will be restricted in norm to ensure that $x_k + s$ belongs to

$$\mathcal{B}_k \stackrel{\text{def}}{=} \{x \in \mathbb{R}^n \mid \|x - x_k\| \leq \Delta_k\}, \tag{3.2}$$

where $\Delta_k$ is a positive parameter. The ball $\mathcal{B}_k$ is called the *trust region*, and $\Delta_k$ is the *trust-region radius.* This radius will be adjusted by the algorithm from iteration to iteration.

In order to keep our analysis simple and hopefully intuitive, we choose a simple set of assumptions.

**A.1** The objective function $f$ is twice continuously differentiable and its gradient and Hessian is uniformly bounded over $\mathbb{R}^n$, which means that there exist constants $\kappa_{\rm fg} > 0$ and $\kappa_{\rm fh} > 0$ such that

$$\|\nabla f(x)\| \leq \kappa_{\rm fg}$$

and

$$\|\nabla^2 f(x)\| \leq \kappa_{\rm fh}$$

for all $x \in \mathbb{R}^n$.

**A.2** The objective function is bounded below on $\mathbb{R}^n$.

**A.3** The Hessians of all models generated are uniformly bounded, that is there exists a constant $\kappa_{\rm mh} > 0$ such that

$$\|H_k\| \leq \kappa_{\rm mh}$$

for all $x \in \mathcal{B}_k$.

In what follows, we will also use the constant $\kappa_{\rm h} = \max[\kappa_{\rm fg}, \kappa_{\rm fh}, \kappa_{\rm mh}]$.

## 3.1 Properties of multivariate interpolation models

We now turn to the determination of the model (3.1). As stated above, we will derive the values of $g_k$ and $H_k$ from a set of *interpolation conditions* of the form

$$m_k(y) = f(y) \;\; \text{for each} \;\; y \in Y, \tag{3.3}$$

for some *interpolation set* $Y$. If a 'fully quadratic' model of the form (3.1) is used, $Y$ needs to contain $p = (n+1)(n+2)/2$ points to determine $g_k$ and $H_k$ uniquely. However, even $p$ interpolation conditions of the form (3.3) are in general not sufficient for this determination if the number of variables, $n$, is larger than one. For instance, six interpolation points on a circle in the plane do not specify a two-dimensional quadratic uniquely, because any quadratic which is a multiple of the equation of the circle can be added to the interpolant without affecting (3.3). One therefore sees that some *geometric conditions* on $Y$ must be added to the conditions (3.3) to ensure the existence and uniqueness of the quadratic interpolant. In our example, we must have that the interpolation points do not lie on any quadratic surface in $\mathbb{R}^n$ or that the chosen basis includes terms of higher degree (see [2] or [23] for details). More formally, we need the condition referred to as poisedness, which relates directly to the interpolation points and the approximation space.

In order to continue the discussion of the geometry of the interpolation set, we need to explore multivariate interpolation techniques a little further, which we do by considering the more general problem of finding interpolating polynomials of degree $d$. To be able to approximate in the desired interpolation space (the quadratic polynomials, say) one typically chooses a suitable basis (for example the monomials) to initiate the process. The techniques then differ in that they combine these basis functions in different ways, although the final result is clearly unique in the interpolation space if the geometry of the interpolation set is suitable. At variance with Powell's proposals, which are based on the use of the Lagrange interpolation polynomials as a tool to build the interpolant, we will, in this paper, emphasize Newton fundamental polynomials. Although there are some significant computational advantages in using Newton polynomials, in the context of our discussion the main reason for this choice is that we will exploit an error formula associated with this latter class of polynomials.

It is interesting to note at this point that the choice of the monomials as an initial basis for the interpolation space has the advantage that any known sparsity structure in the objective function's Hessian may directly be exploited. For instance, if the $(i,j)$-th entry of this Hessian is known to be zero, then the monomial $x_i x_j$ may simply be removed from the initial basis, and the interpolating model will automatically exhibit the desired sparsity pattern in its Hessian. Other choices for the initial basis are nevertheless possible: for instance, [24] considers using Bernstein polynomials for improved numerical stability.

In the framework of multivariate interpolation using Newton's fundamental polynomials, the points $y$ in the interpolation set $Y$ are organized into $d+1$ *blocks* $Y^{[\ell]}$, $(\ell = 0, \ldots, d)$, the $\ell$-th block containing $|Y^{[\ell]}| = \binom{\ell+n-1}{\ell}$ points. To each point $y_i^{[\ell]} \in Y^{[\ell]}$ corresponds a single *Newton fundamental polynomial* of degree $\ell$ satisfying the conditions

$$N_i^{[\ell]}(y_j^{[m]}) = \delta_{ij}\delta_{\ell m} \text{ for all } y_j^{[m]} \in Y^{[m]} \text{ with } m \le \ell. \tag{3.4}$$

For instance, if we consider cubic interpolation on a regular grid in the plane, we require ten interpolation points using four blocks

$$Y^{[0]} = \{(0,0)\}, \quad Y^{[1]} = \{(1,0),(0,1)\},$$

$$Y^{[2]} = \{(2,0),(1,1),(0,2)\} \text{ and } Y^{[3]} = \{(3,0),(2,1),(1,2),(0,3)\},$$

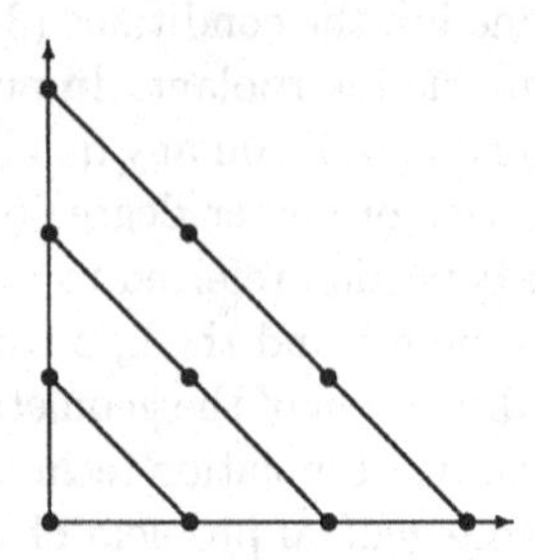

**Figure 1** Interpolation set and the four blocks (connected by thick lines) for cubic interpolation on a regular grid in the plane.

as shown in Figure 1.

The interpolating polynomial $m(x)$ is then given as

$$m(x) = \sum_{y_i^{[\ell]} \in Y} \lambda_i^{[\ell]}(Y, f) N_i^{[\ell]}(x),$$

where the coefficients $\lambda_i^{[\ell]}(Y, f)$ are generalized finite differences applied to $f$. We refer the reader to [23] for more details on these and multivariate interpolation in general.

We will now return to the concept of poisedness. Thus we need to consider in more detail the procedure for constructing the basis of fundamental Newton polynomials as described in [23]. Namely we consider the procedure below for any given $Y$.

**Procedure CNP for constructing fundamental Newton polynomials**

Initialize the $N_i^{[\ell]}$ ($i = 1, \ldots, |Y^{[\ell]}|$, $\ell = 0, \ldots, d$) to the chosen polynomial basis (the monomials).
Set $Y_{temp} = \emptyset$.
For $\ell = 0, \ldots, d$,
  for $i = 1, \ldots, |Y^{[\ell]}|$
    choose some $y_i^{[\ell]} \in Y \setminus Y_{temp}$ such that $|N_i^{[\ell]}(y_i^{[\ell]})| \neq 0$,
    if no such $y_i^{[\ell]}$ exists in $Y \setminus Y_{temp}$, reset $Y = Y_{temp}$ and stop
      (the basis of Newton polynomials is incomplete),
    $Y_{temp} \leftarrow Y_{temp} \cup \{y_i^{[\ell]}\}$,
    normalize the current polynomial by

$$N_i^{[\ell]}(x) \leftarrow N_i^{[\ell]}(x)/|N_i^{[\ell]}(y_i^{[\ell]})|, \tag{3.5}$$

    update all Newton polynomials in block $\ell$ and above by

$$N_j^{[\ell]}(x) \leftarrow N_j^{[\ell]}(x) - N_j^{[\ell]}(y_i^{[\ell]}) N_i^{[\ell]}(x) \quad (j \neq i,\ j = 1, \ldots, |Y^{[\ell]}|), \tag{3.6}$$

$$N_j^{[k]}(x) \leftarrow N_j^{[k]}(x) - N_j^{[k]}(y_i^{[\ell]})N_i^{[\ell]}(x) \quad (j = 1, \ldots, |Y^{[k]}|,\ k = \ell + 1, \ldots, d).$$

end

End (the basis of Newton polynomials is complete).

Clearly, poisedness relates to nonzero pivots in (3.5). Notice that after applying procedure CNP, $Y$ is always poised since we only include the points that create nonzero pivots. This is true even if the procedure stops with an incomplete basis of Newton polynomials, which then results in an interpolating polynomial which is not of full degree $d$ (meaning that it does not include contributions of all the monomials of degree $d$). In practice we need sufficiently large pivots, which is equivalent to 'well-poisedness'. Thus checking that $|N_i^{[\ell]}(y_i^{[\ell]})| \neq 0$ is replaced by $|N_i^{[\ell]}(y_i^{[\ell]})| \geq \theta$, for some $\theta > 0$. We call $\theta$ the pivot threshold. We will show that if throughout the algorithm the interpolation problem can be made sufficiently well-poised we are able to ensure the existence of a bound on the distance between the interpolating polynomial and interpolated function at a point $x$ ($x \notin Y$). Otherwise we provide a mechanism that guarantees we can find a suitable interpolation for which the bound holds. This bound depends upon an important property proved by Sauer and Xu and uses the concept of a *path* between the zero-th block and $x$, which uses a sequence of points of $Y$ of the form

$$\pi(x) = (y_0^{[0]}, y_1^{[1]}, \ldots, y_d^{[d]}, y_{d+1}^{[d+1]} = x)$$

where

$$y_i^{[i]} \in Y^{[i]} \quad (i = 0, \ldots, d).$$

A path therefore contains, besides $x$ itself, exactly one interpolation point in each block. Let us denote by $\Pi(x) = \{\pi(x)\}$, the set of all possible paths from $Y^{[0]}$ to $x$. Using this notion, Sauer and Yu [23] derive in their Theorem 3.11 a bound on $|f(x) - m(x)|$, where $m(x)$ is the polynomial interpolating the function $f(x)$ at the points in $Y$. This bound was further simplified in [25], giving that

$$|f(x) - m(x)| \leq \frac{n^{d+1}\|f^{(d)}\|_\infty}{(d+1)!} \sum_{\pi(x)\in\Pi(x)} \left[\prod_{i=0}^{d} \|y_{i+1}^{[i+1]} - y_i^{[i]}\|_\infty |N_i^{[i]}(y_{i+1}^{[i+1]})|\right], \tag{3.7}$$

for all $x$, where $f^{(d)}$ is the $d$-th derivative of $f$. Interestingly, the quantities $N_i^{[i]}(y_{i+1}^{[i+1]})$ are all computed in the course of the evaluation of the generalized finite differences $\lambda_i^{[\ell]}(Y, f)$. We see that the error between $m(x)$ and $f(x)$ is smaller if we can make the values $N_i^{[i]}(y_{i+1}^{[i+1]})\|y_{i+1}^{[i+1]} - y_i^{[i]}\|_\infty$ small. This fact could possibly be exploited in a specific algorithm for computing the interpolation point $y_i$ in a given domain, but we will here focus on the use of the bound (3.7), and thus of these values, for the purpose of proving convergence of an algorithm based on multivariate polynomial interpolation. We will further verify that, if all the interpolation points and the point $x$ are chosen in a given hypersphere of radius $\delta$, it is then possible to provide an upper bound on the maximal error, which will be of interest in the rest of this paper. The following theorem establishes a relationship between well-poisedness and such a bound.

**Theorem 1** *Suppose we are given $\theta > 0$, $Y \subset \mathcal{Q}_k(\delta)$, where $\mathcal{Q}_k(\delta) \stackrel{\text{def}}{=} \{y \mid \|y-x_k\| \leq \delta\}$ and the basis of Newton polynomials (possibly incomplete) which is built for $Y$ using procedure CNP with $\theta$ as a pivot threshold. Then there exists a number $\kappa_n(\theta) > 0$ such that $|N_j^{[\ell]}(x)| \leq \kappa_n(\theta)$, for any $x \in \mathcal{Q}_k(\delta)$, for $l = 0, \ldots, d$ and $j = 1, \ldots, |Y^{[\ell]}|$.*

**Proof** For simplicity we number the Newton polynomials (and the corresponding points) in this proof consecutively without regard to blocks. We will use an induction argument. We use procedure CNP with the $N_j$, $j = 1, \ldots, |Y|$ initialized by some polynomial basis (e.g. monomials). Thus on the bounded set $\mathcal{Q}_k(\delta)$, $|N_j(x)|$ are all bounded, say by $M$.
*Induction step:* Suppose we have successfully executed procedure CNP through $i$ pivots and there exists $M > 0$ such that $|N_j(x)| < M$ for $j = 1, \ldots, |Y|$ and for $x \in \mathcal{Q}_k(\delta)$. Consider the next step. First we normalize the $i+1$-st polynomial $N_{i+1}$; since we follow the pivot threshold rule, then $|N_{i+1}(x_{i+1})| > \theta$. Thus after the normalizing $|N_{i+1}(x)| < M/\theta$. After updating the other polynomials of the same block as in (3.6) we have $|N_j(x)| < M + M^2/\theta$. Therefore the Newton polynomials are still bounded on $\mathcal{Q}_k(\delta)$.
The induction is complete. The bound depends on the initial $M$, the threshold $\theta$ and the number of Newton polynomials. Since the last is finite, the bound is finite and we set $\kappa_n(\theta)$ to this bound. □

Unfortunately the dependence of $\kappa_n(\theta)$ upon $\theta$ means that it may become arbitrarily large as $\delta$ tends to zero. In practice $\delta$ is bounded away from zero but nevertheless to guarantee convergence we need to provide other means of maintaining the bound. This is the purpose of the next theorem.

**Theorem 2** *Assume that an arbitrary $x_k \in \mathbb{R}^n$ and a $\delta > 0$ are given, together with the interpolation degree $d$. Then it is possible to construct an interpolation set $Y$ yielding a complete basis of Newton polynomials such that all $y \in Y$ satisfy*

$$y \in \mathcal{Q}_k(\delta) \stackrel{\text{def}}{=} \{y \mid \|y - x_k\| \leq \delta\}, \tag{3.8}$$

*and also such that*

$$|N_j^{[\ell]}(x)| \leq 2^{|Y^{[d]}|} \tag{3.9}$$

*for all $\ell = 0, \ldots, d$, all $j = 1, \ldots, |y^{[\ell]}|$ and all $x \in \mathcal{Q}_k(\delta)$.*

**Proof** We consider the procedure CNP but instead of choosing the interpolation points $y_i^{[\ell]}$ from a given set $Y$, we determine them successively by

$$y_i^{[\ell]} = \arg \max_{x \in \mathcal{Q}_k(\delta)} |N_i^{[\ell]}(x)|. \tag{3.10}$$

Note that, because of the independence of the polynomials in the basis, none of the fundamental polynomials can become identically zero, and thus the value of the maximum in (3.10), and thus of the denominator in (3.5), is always nonzero. Applying this algorithm, we see that the normalization step ensures that

$$\max_{x \in \mathcal{Q}_k(\delta)} |N_i^{[\ell]}(x)| \leq 1, \tag{3.11}$$

except that this value may be modified when polynomials of the same block are updated by (3.6) later in the computation. Consider one such update: using (3.6) and (3.11), we have that

$$\begin{aligned} \max_{x \in \mathcal{Q}_k(\delta)} |\hat{N}_j^{[\ell]}(x)| &\leq \max_{x \in \mathcal{Q}_k(\delta)} |N_j^{[\ell]}(x)| + |N_j^{[\ell]}(y_i^{[\ell]})| \max_{x \in \mathcal{Q}_k(\delta)} |N_i^{[\ell]}(x)| \\ &\leq 2 \max_{x \in \mathcal{Q}_k(\delta)} |N_j^{[\ell]}(x)|, \end{aligned}$$

where $\hat{N}_j^{[\ell]}$ denotes the $j$-th polynomial of the $\ell$-th level after the update (3.6). Thus, at the end of the process and taking into account that (3.11) holds for $i = j$ when $N_j^{[\ell]}(x)$ is first normalized,

$$\max_{x \in \mathcal{Q}_k(\delta)} |N_j^{[\ell]}(x)| \leq 2^{|Y^{[\ell]}|-j} \leq 2^{|Y^{[\ell]}|} \leq 2^{|Y^{[d]}|}, \tag{3.12}$$

as desired. □

The number of interpolation points in the last block, $|Y^{[d]}|$, only depends on the problem dimension $n$ and the interpolation degree $d$. Notice that to construct an interpolation model of the form (3.1) we need $x_k \in Y$. This can be guaranteed by choosing $y_1^{[0]} = x_k$, since $N_1^{[0]}$ is a nonzero constant.

We now return to the case of quadratic models of the form (3.1) and define what we mean by an adequate geometry of the interpolation set. In the spirit of Theorem 2, we assume that we are at iteration $k$ of the algorithm, where $x_k$ is known (but arbitrary). We also assume that we are given a set $\mathcal{Q}_k(\delta)$ (as defined in (3.8)) for some $\delta > 0$. We then say that $Y$ is *adequate in* $\mathcal{Q}_k(\delta)$ whenever the cardinality of $Y$ is at least $n+1$ (which means the model is at least fully linear) and

$$y \in \mathcal{Q}_k(\delta) \;\text{ for all }\; y \in Y, \tag{3.13}$$

$$|N_i^{[\ell]}(y_j^{[\ell+1]})| \leq \kappa_{\rm n} \quad (i = 1, \ldots, |Y^{[\ell]}|,\ j = 1, \ldots, |Y^{[\ell+1]}|,\ \ell = 0, \ldots, d-1), \tag{3.14}$$

and

$$|N_i^{[d]}(x)| \leq \kappa_{\rm n} \quad (i = 1, \ldots, |Y^{[d]}|,\ x \in \mathcal{Q}_k(\delta)), \tag{3.15}$$

where $\kappa_{\rm n}$ is any positive constant such that $\kappa_{\rm n} > 2^{|Y^{[d]}|}$. (This choice of $\kappa_{\rm n}$ is merely intended to make (3.14) and (3.15) possible in view of Theorem 2.)

The value of $\kappa_n$ is very large, except for small values of $n$ and $d$, and only serves a theoretical purpose. In practice the threshold pivoting strategy may be a better way to verify the adequacy of the geometry because as long as the new point satisfies the threshold provision we can include it. The use of the procedure suggested in the proof of Theorem 2 may thus be viewed as a last resort if threshold pivoting rules do not prevent the $|N_j^{[\ell]}(y_j^{[\ell+1]})|$ from becoming very large. Interestingly, it is possible to prove that, if this procedure is applied for example to computing a linear model, then pivots cannot be arbitrarily small.

**Lemma 3** *For given $x_k$ and $\delta$ there exist $\theta > 0$ and $Y$ adequate in $\mathcal{Q}_k(\delta)$, which can constructed by CNP with $\theta$ as the pivoting threshold.*

**Proof** We will construct $Y$ exactly as in Theorem 2. We only need to show that the pivot value in that case stays above a certain positive number. We initialize the Newton polynomial basis with the monomials. Then the first pivot equals 1. It is easy to see from the construction process of the basis and from the fact that we start with monomials that at the $i$-th pivot, for $0 < i \leq n$, the $i$-th polynomial is of the form $a_0 + a_1x_1 + \ldots + a_{i-1}x_{i-1} + x_i$. Thus the maximum absolute value of this polynomial in $\mathcal{Q}_k(\delta)$ is at least $\delta$. Therefore by choosing $0 < \theta < \delta$ we guarantee the building of at least a full linear model in $\mathcal{Q}_k(\delta)$. Then from Theorem 2 we know that $Y$ is adequate in $\mathcal{Q}_k(\delta)$. □

Note that one could equally well use higher-order polynomials and obtain a similar results as long as an appropriate relationship between $\theta$ and $\delta$ is chosen. For example if the highest monomial term is $x_i^2$ then one needs $0 < \theta < \min[\delta, \delta^2]/4$.

Unfortunately, this result is of little theoretical use, because it does not imply that, if an interpolation set $Y$ such that $y \in \mathcal{Q}_k(\delta)$ for all $y \in Y$ is determined by the CNP procedure with threshold $\theta$, then it is adequate in $\mathcal{Q}_k(\delta)$. This property would allow us to decide if a model is adequate just by considering the associated pivot values, an algorithmically desirable feature.

If $Y$ is adequate in some ball centred at $x_k$, we may combine (3.7) and Theorem 2 to obtain the following useful result.

**Theorem 4** *Assume that (A.1) and (A.3) hold, that we are using a model of the form (3.1) and that $Y$ is adequate in $\mathcal{Q}_k(\delta)$ defined in (3.8). Then*

$$|f(x) - m_k(x)| \leq \kappa_{em} \max[\delta^2, \delta^3], \tag{3.16}$$

*and*

$$\|\nabla f(x_k) - g_k\| \leq \kappa_{eg} \max[\delta, \delta^2] \tag{3.17}$$

*for all $x \in \mathcal{Q}_k(\delta)$ and some constants $\kappa_{em}, \kappa_{eg} > 0$ independent of $k$.*

**Proof** We consider the bound (3.7) for $d = 1$ and $d = 2$. Because of (A.1) and the definition of $\kappa_h$, we obtain that the quantity $\|f^{(d)}\|$ is bounded above by $\kappa_h$. Furthermore, since $Y$ is adequate in $\mathcal{Q}_k(\delta)$, we may deduce that

$$\|y_i - x_k\| \leq \delta \quad (i = 1, \ldots, p),$$

and therefore that

$$\|y_i - y_j\| \leq 2\delta \quad (i, j = 1, \ldots, p,\ i \neq j).$$

Furthermore, again using the adequacy of $Y$ in $\mathcal{Q}_k(\delta)$, we deduce that (3.14) and (3.15) hold. Combining these bounds with (3.7) for $d = 1$ and $d = 2$ then yields that

$$|f(x) - m_k(x)| \leq 2n^6 \kappa_h \kappa_n \max[\delta^2, \delta^3]$$

which obviously gives (3.16) with $\kappa_{em} \stackrel{\text{def}}{=} 2n^6 \kappa_h \kappa_n$.

If $g_k = \nabla f(x_k)$ then (3.17) follows trivially. Otherwise, since both the model $m_k$ and the objective function are twice continuously differentiable by (A.1) and (A.3), Taylor's theorem gives, for every $h$ satisfying the bound $\|h\| \leq \delta$, that

$$f(x_k + h) = f(x_k) + \langle \nabla f(x_k), h \rangle + \tfrac{1}{2} \langle h, \nabla^2 f(\xi_k) h \rangle$$

and

$$m_k(x_k + h) = f(x_k) + \langle g_k, h \rangle + \tfrac{1}{2} \langle h, H_k h \rangle,$$

where $\xi_k$ belongs to the segment $[x_k, x_k + h]$. Taking the difference of these two equations, we deduce that

$$\langle \nabla f(x_k) - g_k, h \rangle = f(x_k + h) - m_k(x_k + h) - \tfrac{1}{2} \langle h, [\nabla^2 f(\xi_k) - H_k] h \rangle, \quad (3.18)$$

and therefore, using (3.16), (A.1), (A.3) and the Cauchy–Schwarz inequality, that

$$\begin{aligned} |\langle \nabla f(x_k) - g_k, h \rangle| &\leq \kappa_{em} \max[\delta^2, \delta^3] + \tfrac{1}{2} |\langle h, \nabla^2 f(\xi_k) h \rangle| + \tfrac{1}{2} |\langle h, H_k h \rangle| \\ &\leq \kappa_{em} \max[\delta^2, \delta^3] + \kappa_h \delta^2. \end{aligned}$$

Choosing now

$$h = \delta \frac{\nabla f(x_k) - g_k}{\|\nabla f(x_k) - g_k\|},$$

we obtain that

$$\|\nabla f(x_k) - g_k\| \leq \kappa_{em} \max[\delta, \delta^2] + \kappa_h \delta \leq (\kappa_{em} + \kappa_h) \max[\delta, \delta^2].$$

This is (3.17) with $\kappa_{eg} = \kappa_{em} + \kappa_h$. □

Before we state our algorithmic framework, we need to introduce how we consider updating the interpolation set. We will only use two operations on this set: the *insertion* of a given point $x$ in $Y$ and the *improvement* of the model's geometry in a given ball $\mathcal{Q}_k(\delta)$ for some $\delta > 0$.

In the first case (the *insertion* of $x$), we may simply add $x$ to $Y$ if $|Y| < p$, or we need to remove a point of $Y$, if $|Y|$ is already maximal ($|Y| = p$). Ideally, this point should be chosen to make the geometry of $Y$ as good as possible. There are various ways to attempt to achieve this goal. For instance, one might choose to remove $y_i^{[\ell]}$ such that $|N_i^{[\ell]}(x)|$ is maximal, therefore trying to make the pivots as large as possible, but other techniques are possible and will not be discussed here. It is important to note that, since $x$ is given irrespective of the geometry of the points already in $Y$, there is no guarantee that $Y$ will remain adequate in $\mathcal{B}_k$ if it has that property before the insertion.

The second operation (*improvement* of the model's geometry in $\mathcal{Q}_k(\delta)$) promotes making $Y$ adequate in $\mathcal{Q}_k(\delta)$. There is no need to specify exactly how this is done for the purpose of our theory, except that we must guarantee that this property is obtained after a *finite number* of such improvements and that $x_k$ must remain in $Y$. For instance, a reasonable strategy consists in first eliminating a point $y_i^{[\ell]} \in Y$ which is not in $\mathcal{Q}_k(\delta)$ (if such a point exists), and replacing it in $Y$ by a point $z$ which maximizes $N_i^{[\ell]}$ within $\mathcal{Q}_k(\delta)$. If no such exchange is possible, one may then consider replacing interpolation points in $Y \setminus \{x_k\}$ by new ones in order to ensure that (3.14) and (3.15) hold after a finite number of replacements. The proof of Theorem 2 guarantees that this is indeed possible whereas Theorem 1 attempts to do the same more efficiently. We refer the reader to [21], [6] or [22] for more details on practical procedures for improving the model's geometry.

## 3.2 The algorithmic framework

We now describe our class of algorithms formally as follows. We assume that the constants

$$0 < \eta_0 \leq \eta_1 < 1, \quad 0 < \gamma_0 \leq \gamma_1 < 1 \leq \gamma_2, \quad \epsilon_g > 0 \text{ and } 1 \leq \mu$$

are given. We also use the notation

$$\mathcal{A}_k = \{y \in \mathbb{R}^n \mid \|y - x_k\| \leq \mu \|g_k\|\}.$$

**Derivative-free trust-region algorithm**

**Step 0: Initialization**

Let $x_s$ and $f(x_s)$ be given. Choose an initial interpolation set $Y$ containing $x_s$ and at least one other point. Then determine $x_0 \in Y$ such that $f(x_0) = \min_{y_i \in Y} f(y_i)$. Choose an initial trust-region radius $\Delta_0 > 0$. Set $k = 0$.

**Step 1: Build the model**

Using the interpolation set $Y$, build a model $m_k(x_k + s)$, possibly restricting $Y$ to a subset containing $x_k$, such that conditions (3.3) hold for the resulting $Y$. If $\|g_k\| \leq \epsilon_g$ and $Y$ is inadequate in $\mathcal{A}_k$, then *improve* the geometry until $Y$ is adequate in $\mathcal{Q}_k(\delta_k)$ for some $\delta_k \in (0, \mu\|g_k\|]$.

**Step 2: Minimize the model within the trust region**

Compute the point $x_k + s_k$ such that

$$m_k(x_k + s_k) = \min_{x \in \mathcal{B}_k} m_k(x). \tag{3.19}$$

Compute $f(x_k + s_k)$ and the ratio

$$\rho_k \stackrel{\text{def}}{=} \frac{f(x_k) - f(x_k + s_k)}{m_k(x_k) - m_k(x_k + s_k)}. \tag{3.20}$$

**Step 3: Update the interpolation set**

- If $\rho_k \geq \eta_1$, *insert* $x_k + s_k$ in $Y$, dropping one of the existing interpolation points if $|Y| = p$.
- If $\rho_k < \eta_1$ and $Y$ is inadequate in $\mathcal{B}_k$, *improve* the geometry in $\mathcal{B}_k$.

**Step 4: Update the trust-region radius**

- If $\rho_k \geq \eta_1$, then set

$$\Delta_{k+1} \in [\Delta_k, \gamma_2 \Delta_k]. \tag{3.21}$$

- If $\rho_k < \eta_1$ and $Y$ is adequate in $\mathcal{B}_k$, then set

$$\Delta_{k+1} \in [\gamma_0 \Delta_k, \gamma_1 \Delta_k]. \tag{3.22}$$

- Otherwise, set $\Delta_{k+1} = \Delta_k$.

**Step 5: Update the current iterate**

Determine $\hat{x}_k$ such that

$$f(\hat{x}_k) = \min_{\substack{y_i \in Y \\ y_i \neq x_k}} f(y_i). \tag{3.23}$$

Then, if

$$\hat{\rho}_k \stackrel{\text{def}}{=} \frac{f(x_k) - f(\hat{x}_k)}{m_k(x_k) - m_k(x_k + s_k)} \geq \eta_0, \tag{3.24}$$

set $x_{k+1} = \hat{x}_k$. Otherwise, set $x_{k+1} = x_k$. Increment $k$ by one and go to Step 1.

**End of algorithm**

If the algorithm is able to progress, that is if (3.24) holds, we say that the iteration is *successful*. If (3.24) fails, then the iteration is *unsuccessful*. We denote the index set of all successful iterations by $\mathcal{S}$. In other words,

$$\mathcal{S} = \{k \mid \hat{\rho}_k \geq \eta_0\}. \tag{3.25}$$

We also define $\mathcal{R}$ to be the index set of all iterations where the trust-region radius is reduced, that is

$$\mathcal{R} = \{k \mid \Delta_{k+1} < \Delta_k\}. \tag{3.26}$$

This description calls for several comments.

1. Our algorithmic specification is admittedly simplistic. It is enough to consider the methods proposed in [21], [6] or [22] to be convinced that practical algorithms involve a number of additional features that enhance efficiency. We recall here that our purpose is merely to explore a framework in which convergence theory can be discussed, and the above will serve that goal.

2. Any further improvement in the model, beyond what the algorithm explicitly includes, is also possible. For instance, one might wish to include $x_k + s_k$ in $Y$, even if $\rho_k < \eta_1$, provided it does not worsen the quality of the model. Indeed, we have computed $f(x_k+s_k)$ and any such evaluation should be exploited if at all possible. One could also decide to perform a model improvement step if $\rho_k$ is very small, indicating a bad fit of the model to the objective function. Or one could require that the model remains at least linear for all $k$. Any further decrease in function values obtained within these steps is then taken into account by Step 5.

3. The loop in Step 1 may be viewed as an inner iteration for improving the model's geometry, whose purpose is to ensure that $g_k$, the first-order information for the model, is not too far from the true first-order information for the objective function. This feature is close in spirit to a proposal of [5]. It remains to be seen whether this inner iteration can be well-defined and finite.

   Other ways to ensure that the model's gradient is correctly approximated when it becomes small can of course be considered.

4. The trust-region management is merely outlined in our description, restricting a usually more involved procedure to its essentials.

5. We have not mentioned a stopping test in the algorithm, which is acceptable for studying convergence properties, but practical methods would require a stopping criterion. One possible test is to stop the calculation if either the trust-region radius falls below a certain threshold, or the model's gradient becomes sufficiently small and the geometry of the interpolation set is adequate.

6. We have assumed that the models are at most quadratic, but this need not be so in practical implementations. For instance, [6] suggests the use of models of degree exceeding two and the theory can be readily extended to account for this. Models that are less than fully quadratic, as is allowed by our framework, are also beneficial when function evaluations are expensive since we are able to use them as soon as they are available.

7. The set $\mathcal{A}_k$ plays a role similar to that of the trust region $\mathcal{B}_k$, but is intended to monitor the first-order part of the model. The use of these two regions is reminiscent of the two regions proposed in [22]. Note that the additional computational effort required to obtain an adequate geometry may be practically adjusted by a suitable choice of $\epsilon_g$.

We conclude this section by stating four useful properties of our algorithm.

**Lemma 5**
*(i) For all $k$, if $\rho_k \geq \eta_1$, then $\hat{\rho}_k \geq \eta_0$ and thus iteration $k$ is successful.*

*(ii) If $k \in \mathcal{R}$, then $Y$ is adequate in $\mathcal{B}_k$.*

*(iii) The conditions of Step 1 may be satisfied in a finite number of improvements of the geometry, unless $\nabla f(x_k) = 0$.*

*(iv) There can only be a finite number of iterations such that $\rho_k < \eta_1$ before the trust-region radius is reduced in (3.22).*

**Proof** If $\rho_k \geq \eta_1$ the mechanism of Step 3 implies that $x_k + s_k$ is added to the interpolation set $Y$. Hence

$$f(\hat{x}_k) \leq f(x_k + s_k)$$

because of the definition of $\hat{x}_k$ in (3.23). Thus

$$\hat{\rho}_k = \frac{f(x_k) - f(\hat{x}_k)}{m_k(x_k) - m_k(x_k + s_k)} \geq \frac{f(x_k) - f(x_k + s_k)}{m_k(x_k) - m_k(x_k + s_k)} = \rho_k \geq \eta_1 \geq \eta_0,$$

and conclusion (i) follows. Property (ii) immediately results from the definition of $\mathcal{R}$ and the mechanism of Step 4. Property (iv) follows from our definition of the model's improvement in $\mathcal{B}_k = \mathcal{Q}_k(\Delta_k)$. We must be more careful for (iii) because of the interaction between $g_k$ and $\delta_k$. Consider the following procedure. First choose a scalar $\nu \in (0, 1)$, set $g_k^{(0)} = g_k$ and determine $Y$ such that it is adequate in $\mathcal{Q}_k(\nu\epsilon_g)$ (in a finite number of improvements of the geometry). If the resulting $g_k^{(1)}$ satisfies $\mu\|g_k^{(1)}\| \geq \nu\epsilon_g$, the procedure stops with $\delta_k = \nu\epsilon_g$. Otherwise, that is if $\mu\|g_k^{(1)}\| < \nu\epsilon_g$, determine $Y$ such that it is adequate in $\mathcal{Q}_k(\nu^2\epsilon_g)$. Then again either the procedure stops or the radius of $\mathcal{Q}_k$ is again multiplied by $\nu$, and so on. The only possibility for this procedure to be infinite (and to require an infinite number of improvements of the geometry) is if $\mu\|g_k^{(i)}\| < \nu^i\epsilon_g$ for all $i \geq 0$. But $Y$ is adequate in each of the $\mathcal{Q}_k(\nu^i\epsilon_g)$ and therefore Theorem 4 implies that

$$\|\nabla f(x_k) - g_k^{(i)}\| \leq \kappa eg \nu^i \epsilon_g$$

for each $i \geq 0$, and thus, using the triangle inequality, that, for all $i \geq 0$,

$$\|\nabla f(x_k)\| \leq \|g_k^{(i)}\| + \|\nabla f(x_k) - g_k^{(i)}\| \leq \left(\frac{1}{\mu} + \kappa eg\right) \nu^i \epsilon_g.$$

Since $\nu \in (0,1)$, this implies that $\nabla f(x_k) = 0$, as desired. □

The first of these results indicates that the iteration has to be successful if the model fits the objective function well enough. The second says that iterations when the trust region is reduced correspond to adequate geometry. The third property indicates that if an infinite iteration occurs within Step 1, this must be because the current iterate is critical, which would then solve our problem. Note that in this case the number of successful iterations is finite. The fourth guarantees the overall coherence of the method in that it reduces the trust-region radius eventually, if improvements in the model's geometry are not enough to ensure progress.

# 4 A brief convergence analysis

In many ways, our algorithm shares the common features of all trust-region methods: the use of a model of the objective function, the role of the ratio $\rho_k$ of achieved versus predicted reduction, most of the trust-region updating technique. However, it differs from this standard framework in three important aspects. The first is that the model and objective function gradients need not coincide. Indeed, the objective function gradient is unavailable and that of the model is derived from the interpolation conditions (3.3) which are not designed to yield a good approximation of the gradient at the current iterate (as would a technique using finite difference approximations). We thus have to verify the coherence of the model's gradient with that (unknown) of the objective function. The second important difference from a more traditional trust-region algorithm is that the model is subject to a condition on the geometry of the interpolation points: for the model to be 'valid' this condition must be satisfied. This implies that the model cannot be judged locally inappropriate (and, consequently, the trust-region radius reduced) before the geometry of the interpolation set is itself satisfactory. Hence the additional condition in the second part of Step 4. A third difference is also worth indicating: because new interpolation points may be computed in the geometry improvement phase of Steps 1 and 3, it is possible that the value of the objective at one of these points is lower than that of the current iterate. Since function values are assumed to be expensive, this lower value is 'exploited' by shifting the current iterate to the point with lowest value (in Step 5), which means that the algorithm may generate a new iterate, even if the model is bad (in the sense that $\rho_k$ is small). These three differences are unfortunately enough to make the standard convergence theory inapplicable to our framework, and we have to reconsider it in detail. This is the object of this section.

**Lemma 6** *At every iteration $k$, one has*

$$m_k(x_k) - m_k(x_k + s_k) \geq \kappa_{\rm mdc} \|g_k\| \min \left[ \frac{\|g_k\|}{\kappa_{\rm h}}, \Delta_k \right] \tag{4.1}$$

*for some constant $\kappa_{\rm mdc} \in (0, 1)$ independent of $k$.*

**Proof** We do not prove this result in detail, as it is classical in the trust-region literature. It results from the fact that the value of the model at $x_k + s_k$ must be lower than that at the *Cauchy point*, which is defined as the minimizer of the model on the intersection of its steepest descent direction and the trust region $\mathcal{B}_k$. A proof that the bound (4.1) holds at the Cauchy point may be found in [12]. □

We may conclude from Theorem 4 that the error between the objective and the model decreases at least quadratically with the trust-region radius. The smaller this radius becomes, the better the model approximates the objective, which should intuitively guarantee that minimizing the model within the trust region will also decrease the objective function, as desired. We next show that this intuition is vindicated, in that an iteration must be successful if the current iterate is not critical and the trust-region radius is small enough.

**Lemma 7** *Assume that (A.1)–(A.3) hold. Assume furthermore that there exists a constant $\kappa_{\rm g} > 0$ such that $\|g_k\| \geq \kappa_{\rm g}$ for all $k$. Then there is a constant $\kappa_{\rm d} > 0$ such that*

$$\Delta_k > \kappa_{\rm d} \tag{4.2}$$

*for all $k$.*

**Proof** Assume that $k$ is the first $k$ such that

$$\Delta_{k+1} \leq \gamma_0 \min \left[ 1, \frac{\kappa_{\rm mdc} \kappa_{\rm g} (1 - \eta_1)}{\max[\kappa_{\rm h}, \kappa_{\rm em}]} \right]. \tag{4.3}$$

We first note that the mechanism of the algorithm then implies that $k \in \mathcal{R}$, and also, from (3.22), that

$$\Delta_k \leq \min \left[ 1, \frac{\kappa_{\rm mdc} \kappa_{\rm g} (1 - \eta_1)}{\max[\kappa_{\rm h}, \kappa_{\rm cm}]} \right]. \tag{4.4}$$

Observe now that the condition $\eta_1 \in (0, 1)$ and the inequality $\kappa_{\rm mdc} < 1$ imply

$$\kappa_{\rm mdc}(1 - \eta_1) \leq 1. \tag{4.5}$$

Thus condition (4.4) and our assumption on $g_k$ imply

$$\Delta_k \leq \frac{\|g_k\|}{\kappa_{\rm h}}.$$

As a consequence, Lemma 6 immediately gives

$$m_k(x_k) - m_k(x_k + s_k) \geq \kappa_{\rm mdc} \|g_k\| \min\left[\frac{\|g_k\|}{\kappa_{\rm h}}, \Delta_k\right] = \kappa_{\rm mdc} \|g_k\| \Delta_k. \quad (4.6)$$

On the other hand, we may apply Theorem 4 with $\mathcal{B}_k = \mathcal{Q}_k(\Delta_k)$, since $k \in \mathcal{R}$, and thus, by Lemma 5, $Y$ is adequate in $\mathcal{B}_k$. We therefore deduce from (4.6), (3.16), (4.5) and (4.4) that

$$\begin{aligned} |\rho_k - 1| &= \left|\frac{f(x_k + s_k) - m_k(x_k + s_k)}{m_k(x_k) - m_k(x_k + s_k)}\right| \leq \frac{\kappa_{\rm em}}{\kappa_{\rm mdc}\|g_k\|} \max[\Delta_k^2, \Delta_k] \\ &\leq \frac{\kappa_{\rm em}}{\kappa_{\rm mdc}\|g_k\|} \Delta_k \leq 1 - \eta_1 \end{aligned}$$

where we have used (4.4) to deduce the penultimate inequality. Therefore, $\rho_k \geq \eta_1$. The mechanism of Step 4 then ensures that (3.21) is applied, which contradicts the fact that $k \in \mathcal{R}$. Hence there is no $k$ such that (4.3) holds, which gives the desired conclusion with

$$\kappa_{\rm d} = \gamma_0 \min\left[1, \frac{\kappa_{\rm mdc}\kappa_{\rm g}(1 - \eta_1)}{\max[\kappa_{\rm h}, \kappa_{\rm em}]}\right].$$

□

We therefore see that the radius cannot become too small as long as a critical point is not approached. This property is crucial in that it ensures that progress of the algorithm is always possible (except at critical points).

The above results are sufficient to analyze the criticality of the unique limit point of the sequence of iterates when there are only finitely many successful iterations.

**Theorem 8** *Assume that (A.1)–(A.3) hold. Assume furthermore that there are only finitely many successful iterations. Then $x_k = x_*$ for $k$ sufficiently large and $\nabla f(x_*) = 0$.*

**Proof** If an infinite loop occurs within Step 1, the result follows from Lemma 5(iii). Otherwise, the mechanism of the algorithm ensures that $x_* = x_{k_0+1} = x_{k_0+j}$ for all $j > 0$, where $k_0$ is the index of the last successful iteration. Moreover, since all iterations are unsuccessful for sufficiently large $k$, we obtain from Lemma 5(i) that $\rho_k < \eta_1$, and (3.22) along with Lemma 5(iv) thus implies that the sequence $\{\Delta_k\}$ converges to zero and, hence, that $\mathcal{R}$ contains at least one infinite subsequence $\{k_j\}$. Moreover, Lemma 5(ii) implies that $Y$ is adequate in $\mathcal{B}_{k_j}$ for all $j$. Assume now that $\nabla f(x_*) \neq 0$. Using the second part of Theorem 4 have that, for $k > k_0$,

$$\|\nabla f(x_*) - g_{k_j}\| \leq \kappa_{\rm eg} \max[\Delta_k, \Delta_k^2],$$

which, since $\{\Delta_k\}$ converges to zero, then implies

$$\|g_{k_j}\| \geq \kappa_{\rm g} = \tfrac{1}{2}\|\nabla f(x_*)\| > 0$$

for $k \geq k_0$ sufficiently large. But Lemma 7 states that this situation is impossible, and hence $\nabla f(x_*) = 0$, as desired. □

Having proved the desired convergence property for the case where $\mathcal{S}$ is finite, we now restrict our attention to the case where there are infinitely many successful iterations. In this context, we start by proving that at least one accumulation point of the sequence of iterates (when it is infinite) must be critical. The intuition behind this result is that, if $\Delta_k$ is small enough, then the model should approximate the function well, due to Theorem 4, but at the same time, $\Delta_k$ cannot become so small as to prevent progress, as shown by Lemma 7: hence convergence should occur.

**Theorem 9** *Assume that (A.1)–(A.3) hold. Then*

$$\liminf_{k \to \infty} \|g_k\| = 0. \tag{4.7}$$

**Proof** Assume, for a contradiction, that, for all $k$,

$$\|g_k\| \geq \kappa_g \tag{4.8}$$

for some $\kappa_g > 0$ and consider a successful iteration of index $k$. The fact that $k \in \mathcal{S}$ and conditions (4.1), (4.8) and (4.2) and (A.1) then give

$$f(x_k) - f(x_{k+1}) \geq \eta_0[m_k(x_k) - m_k(x_k + s_k)] \geq \kappa_{\rm mdc}\kappa_g\eta_0 \min\left[\frac{\kappa_g}{\kappa_h}, \kappa_d\right].$$

Summing now over all successful iterations from 0 to $k$, we obtain that

$$f(x_0) - f(x_{k+1}) \geq \sum_{\substack{j=0 \\ j \in \mathcal{S}}}^{k} [f(x_j) - f(x_{j+1})] \geq \sigma_k \kappa_{\rm mdc}\kappa_g\eta_0 \min\left[\frac{\kappa_g}{\kappa_h}, \kappa_d\right],$$

where $\sigma_k$ is the number of successful iterations up to iteration $k$. But since there are infinitely many such iterations, we have

$$\lim_{k \to \infty} \sigma_k = +\infty,$$

and the difference between $f(x_0)$ and $f(x_{k+1})$ is unbounded, which clearly contradicts the fact that, according to (A.2), the objective function is bounded below. Our assumption (4.8) must therefore be false, which yields (4.7). □

**Lemma 10** *Assume that (A.1)–(A.3) hold and that $\{k_i\}$ is a subsequence such that*

$$\lim_{i \to \infty} \|g_{k_i}\| = 0. \tag{4.9}$$

*Then*

$$\lim_{i \to \infty} \|\nabla f(x_{k_i})\| = 0. \tag{4.10}$$

**Proof** By (4.9), $\|g_{k_i}\| \leq \epsilon_g$ for sufficiently large $i$, Lemma 5(iii) and the mechanism of Step 1 ensure that $Y$ is adequate in $\mathcal{Q}_{k_i}(\delta_{k_i})$ for sufficiently large $i$, where

$$\delta_{k_i} \leq \mu \|g_{k_i}\|. \tag{4.11}$$

Theorem 4 then allows us to deduce that, for sufficiently large $i$,

$$\|\nabla f(x_{k_i}) - g_{k_i}\| \leq \kappa eg \max[\delta_{k_i}, \delta_{k_i}^2] \leq \kappa eg \max[\mu\|g_{k_i}\|, \mu^2\|g_{k_i}\|^2] \leq \kappa eg\mu\|g_{k_i}\|,$$

where the middle inequality results from (4.11). Hence, , for sufficiently large $i$,

$$\|\nabla f(x_{k_i})\| \leq \|g_{k_i}\| + \|\nabla f(x_{k_i}) - g_{k_i}\| \leq (1 + \kappa eg\mu)\|g_{k_i}\|.$$

The limit (4.9) and this last bound then give (4.10) □

Using this lemma, we obtain the following global convergence result.

**Theorem 11** *Assume that (A.1)–(A.3) hold. Then there is at least one subsequence of successful iterates $\{x_k\}$ whose limit is a critical point, that is*

$$\liminf_{k\to\infty} \|\nabla f(x_k)\| = 0.$$

**Proof** There is no loss of generality in restricting our attention to successful iterations. This result then immediately follows Theorem 9 and Lemma 10. □

Note that the critical limit of the subsequence may be at infinity.

This result corresponds to the usual first step in the convergence analysis of a trust-region algorithm. We now prove that *all* limit points of the sequence of iterates are critical.

**Theorem 12** *Assume that (A.1)–(A.3) hold. Then every limit point $x_*$ of the sequence $\{x_k\}$ is critical, that is $\nabla f(x_*) = 0$.*

**Proof** Assume, for a contradiction, that there is a subsequence of successful iterates, indexed by $\{t_i\} \subseteq \mathcal{S}$, such that

$$\|\nabla f(x_{t_i})\| \geq \epsilon_0 > 0 \tag{4.12}$$

for some $\epsilon_0 > 0$ and for all $i$. Then, because of Lemma 10, we obtain

$$\|g_{t_i}\| \geq 2\epsilon > 0 \tag{4.13}$$

for some $\epsilon > 0$ and for all sufficiently large $i$. Without loss of generality, we may assume that

$$(2 + \kappa_{eg}\mu)\epsilon \leq \tfrac{1}{2}\epsilon_0. \tag{4.14}$$

Theorem 9 then ensures the existence, for each $t_i$, of a first successful iteration $\ell(t_i) > t_i$ such that $\|g_{\ell(t_i)}\| < \epsilon$. Denoting $\ell_i \stackrel{\text{def}}{=} \ell(t_i)$, we thus obtain that there exists another subsequence of $\mathcal{S}$ indexed by $\{\ell_i\}$ such that

$$\|g_k\| \geq \epsilon \text{ for } t_i \leq k < \ell_i \text{ and } \|g_{\ell_i}\| < \epsilon, \tag{4.15}$$

for sufficiently large $k$. We now restrict our attention to the subsequence of successful iterations whose indices are in the set

$$\mathcal{K} \stackrel{\text{def}}{=} \{k \in \mathcal{S} \mid t_i \leq k < \ell_i\},$$

where $t_i$ and $\ell_i$ belong to the two subsequences defined above. Using (4.1), the fact that $\mathcal{K} \subseteq \mathcal{S}$ and (4.15), we deduce that, for $k \in \mathcal{K}$,

$$f(x_k) - f(x_{k+1}) \geq \eta_0[m_k(x_k) - m_k(x_k + s_k)] \geq \kappa_{\text{mdc}}\epsilon\eta_0 \min\left[\frac{\epsilon}{\kappa_{\text{h}}}, \Delta_k\right]. \quad (4.16)$$

But the sequence $\{f(x_k)\}$ is monotonically decreasing and bounded below because of (A.2). Hence it is convergent and the left-hand side of (4.16) must tend to zero when $k$ tends to infinity. This gives

$$\lim_{\substack{k\to\infty \\ k\in\mathcal{K}}} \Delta_k = 0.$$

As a consequence, the second term dominates in the minimum of (4.16) and we obtain, for sufficiently large $k \in \mathcal{K}$,

$$\Delta_k \leq \frac{1}{\kappa_{\text{mdc}}\epsilon\eta_0}[f(x_k) - f(x_{k+1})].$$

We then deduce from this bound that, for sufficiently large $i$,

$$\|x_{t_i} - x_{\ell_i}\| \leq \sum_{\substack{j=t_i \\ j\in\mathcal{K}}}^{\ell_i-1} \|x_j - x_{j+1}\| \leq \sum_{\substack{j=t_i \\ j\in\mathcal{K}}}^{\ell_i-1} \Delta_j \leq \frac{1}{\kappa_{\text{mdc}}\epsilon\eta_0}[f(x_{t_i}) - f(x_{\ell_i})].$$

Using (A.2) and the monotonicity of the sequence $\{f(x_k)\}$ again, we see that the right-hand side of this inequality must converge to zero, and we therefore obtain

$$\lim_{i\to\infty} \|x_{t_i} - x_{\ell_i}\| = 0. \quad (4.17)$$

Now

$$\|\nabla f(x_{t_i})\| \leq \|\nabla f(x_{t_i}) - \nabla f(x_{\ell_i})\| + \|\nabla f(x_{\ell_i}) - g_{\ell_i}\| + \|g_{\ell_i}\|.$$

The first term of the right-hand side tends to zero because of the continuity of the gradient of $f$ (A.1), and is thus bounded by $\epsilon$ for sufficiently large $i$. The third term is bounded by $\epsilon$ by (4.15). Using a reasoning similar to that of Lemma 10, we also deduce that the second term is bounded by $\kappa_{\text{eg}}\mu\epsilon$ for sufficiently large $i$. As a consequence, we obtain from these bounds and (4.14) that

$$\|\nabla f(x_{t_i})\| \leq (2 + \kappa_{\text{eg}}\mu)\epsilon \leq \tfrac{1}{2}\epsilon_0$$

for large enough $i$, which contradicts (4.12). Hence our initial assumption must be false and the theorem follows. □

# 5 Conclusions and perspectives

We have presented a broad class of derivative-free trust-region methods for unconstrained optimization, for which global convergence to first-order critical points can be proved. The definition of this class is inspired by the proposals of [21], [6] and [22]. The derivation of the convergence result is based on a bound on the interpolation error provided by [23].

It is clear to the authors that much additional work remains to specify which members of the class provide the best numerical performance. The class itself could also be extended, and we have provided suggestions for potentially useful generalizations. Finally, a challenging question is to examine whether other bounds than that of Sauer and Xu could be used to derive similar convergence results.

# Acknowledgements

The authors are pleased to acknowledge the helpful and friendly comments made by T. Sauer and C. de Boor on the use of Newton's polynomials. The second author was supported in part by an IBM Cooperative Fellowship. Thanks are also due to M. Roma for his kind support.

# References

[1] Bongartz, I., A.R. Conn, N.I.M. Gould, and Ph.L. Toint (1995) "CUTE: Constrained and Unconstrained Testing Environment", *ACM Trans. Math. Software* **21**, 123–160.

[2] de Boor, C. and A. Ron (1992) "Computational aspects of polynomial interpolation in several variables", *Maths Comput.* **58**, 705–727.

[3] Brent, R.P. (1973) *Algorithms for Minimization Without Derivatives,* Prentice-Hall (Engelwood Cliffs, New Jersey).

[4] Callier, F.M. and Ph.L. Toint (1977) "Recent results on the accelerating property of an algorithm for function minimization without calculating derivatives", in *Survey of Mathematical Programming* (A. Prekopa, ed.), Publishing House of the Hungarian Academy of Sciences, 369–376.

[5] Carter, R.G. (1991) "On the global convergence of trust region methods using inexact gradient information", *SIAM J. Num. Anal.* **28**, 251–265.

[6] Conn, A.R. and Ph.L. Toint (1996) "An algorithm using quadratic interpolation for unconstrained derivative free optimization", in *Nonlinear Optimization and Applications* (G. Di Pillo and F. Gianessi, eds), Plenum Publishing (New York), 27–47.

[7] Dennis. J.E. and R.B. Schnabel (1983), *Numerical Methods for Unconstrained Optimization and Nonlinear Equations,* Prentice-Hall (Englewood Cliffs, New Jersey).

[8] Dennis, J.E. and V. Torczon (1991), "Direct search methods on parallel machines", *SIAM J. Optimization* **1**, 448–474.

[9] Griewank, A. and G. Corliss (1991), *Automatic Differentiation of Algorithms,* SIAM (Philadelphia).

[10] Griewank, A., "Computational differentiation and optimization", in *Mathematical Programming: State of the Art 1994* (J.R. Birge and K.G. Murty, eds), University of Michigan, 102–131.

[11] Harwell Subroutine Library (1995) *A Catalogue of Subroutines (Release 12),* Advanced Computing Department, Harwell Laboratory (Harwell, UK).

[12] Moré, J.J. (1993) "Recent developments in algorithms and software for trust region methods", in *Mathematical Programming: The State of the Art* (A. Bachem, M. Grötschel and B. Korte, eds), Springer Verlag (Berlin), 258–287.

[13] Nelder, J.A. and R. Mead (1965) "A simplex method for function minimization", *Computer J.* **7**, 308–313.

[14] Powell, M.J.D. (1964) "An efficient method for finding the minimum of a function of several variables without calculating derivatives", *Computer J.* **7**, 155–162.

[15] Powell, M.J.D. (1965) "A method for minimizing a sum of squares of non-linear functions without calculating derivatives", *Computer J.* **7**, 303–307, 1965.

[16] Powell, M.J.D. (1970), "A hybrid method for nonlinear equations" in *Numerical Methods for Nonlinear Algebraic Equations* (P. Rabinowitz, editor), Gordon and Breach (London), 87–114.

[17] Powell, M.J.D. (1970) "A Fortran subroutine for solving systems of non-linear algebraic equations", in *Numerical Methods for Nonlinear Algebraic Equations* (P. Rabinowitz, ed.), Gordon and Breach (London), 115–161.

[18] Powell, M.J.D. (1974) "Unconstrained minimization algorithms without computation of derivatives", *Bollettino della Unione Matematica Italiana* **9**, 60–69.

[19] Powell, M.J.D. (1975) "A view of unconstrained minimization algorithms that do not require derivatives", *ACM Trans. Math. Software* **1**, 97–107.

[20] Powell, M.J.D. (1994) "A direct search optimization method that models the objective and constraint functions by linear interpolation", in *Advances in Optimization and Numerical Analysis* (S. Gomez and J-P. Hennart, eds), Kluwer Academic Publishers (Dordrecht), 51–67.

[21] Powell, M.J.D. (1994) "A direct search optimization method that models the objective by quadratic interpolation", Presentation at the 5th Stockholm Optimization Days.

[22] Powell, M.J.D. (1996) "Trust region methods that employ quadratic interpolation to the objective function", Presentation at the 5th SIAM Conference on Optimization.

[23] Sauer, Th. and Y. Xu (1995) "On multivariate Lagrange interpolation", *Maths Comput.* **64**, 1147–1170.

[24] Sauer, Th. (1995) "Computational aspects of multivariate polynomial interpolation", *Adv. Comput. Maths* **3**, 219–238.

[25] Sauer, Th. (1996) "Notes on polynomial interpolation", Private communication.

[26] Toint, Ph.L. (1978) *Unconstrained Optimization: The Analysis of Conjugate Directions Method Without Derivatives and a New Sparse Quasi-Newton Update,* Ph.D. thesis, Department of Mathematics, FUNDP, Namur, Belgium.

[27] Toint, Ph.L. and F.M. Callier (1978) "On the accelerating property of an algorithm for function minimization without calculating derivatives", *J. Optimization Theory Applics* **23**, 531–547. See also same journal **26** (1978), 465–467.

[28] Torczon, V. (1991), "On the convergence of the multidirectional search algorithm", *SIAM J. Optimization* **1**, 123–145.

# Least Squares Fitting to Univariate Data Subject to Restrictions on the Signs of the Second Differences

*I.C. Demetriou*

Department of Economics, University of Athens, 8 Pesmazoglou Street, Athens 10559, Greece

*M.J.D. Powell*

Department of Applied Mathematics and Theoretical Physics, University of Cambridge, Cambridge CB3 9EW, England

**Abstract**

If the data $(x_i, y_i) \in \mathcal{R}^2$, $i=1,2,\ldots,n$, include substantial random errors, then the number of sign changes in the sequence of second differences $y[x_{i-1}, x_i, x_{i+1}]$, $i=2,3,\ldots,n-1$, may be unacceptably large, where we are assuming $x_1 < x_2 < \cdots < x_n$. Therefore we address the problem of calculating the fit $\phi_i$, $i = 1,2,\ldots,n$, that minimizes $\sum_{i=1}^{n}(y_i - \phi_i)^2$, subject to the condition that there are at most $k-1$ sign changes in the sequence $\phi[x_{i-1}, x_i, x_{i+1}]$, $i=2,3,\ldots,n-1$, where $k$ is a prescribed positive integer. It is proved that there exists a partitioning of the data into $k$ disjoint subsets, such that the required fit can be calculated by solving a convex quadratic programming problem on each subset. Further, the piecewise linear interpolant to the fit is either convex or concave on each partition, and there is alternation between the convex and concave partitions. Thus, about twelve years ago, the authors developed some dynamic programming methods that calculate the required fit efficiently, assuming that, if the right hand end of a partition is available, then the left hand end can be calculated without using the data beyond the right hand end. The present paper, however, gives a counter-example to this conjecture. Then a new dynamic programming algorithm is proposed that does not depend on any assumptions, and whose total work should not exceed $\mathcal{O}(kn^3)$ for all values of $k$, but the work before was only about $\mathcal{O}(kn^2)$.

## 1 Introduction

We address the problem of smoothing measured values of a function of one variable. Specifically, the data are the coordinates $(x_i, y_i) \in \mathcal{R}^2$, $i=1,2,\ldots,n$, where the numbers $x_i$, $i = 1,2,\ldots,n$, increase strictly monotonically, and where $y_i$ is the measurement of the underlying function at the point $x_i$. It is usually possible to estimate the magnitude of random errors in the data by forming a table of divided differences, as suggested in Section 4.9 of Hildebrand (1956), for instance. We give particular attention to the second differences

$$y[x_{i-1},x_i,x_{i+1}] = \frac{y_{i-1}}{(x_{i-1}-x_i)\,(x_{i-1}-x_{i+1})} + \frac{y_i}{(x_i-x_{i-1})\,(x_i-x_{i+1})} + \frac{y_{i+1}}{(x_{i+1}-x_{i-1})\,(x_{i+1}-x_i)}, \qquad i=2,3,\ldots,n-1, \tag{1.1}$$

noting that, in this definition, the data $y_{i-1}$, $y_i$ and $y_{i+1}$ are divided by positive, negative and positive numbers, respectively. Thus large random errors tend to cause many sign changes in the sequence $y[x_{i-1},x_i,x_{i+1}]$, $i=2,3,\ldots,n-1$. On the other hand, if the data are exact values of a function $y(x)$, $x_1 \le x \le x_n$, that has a continuous second derivative, then, for $i=2,3,\ldots,n-1$, there exists $\xi_i \in [x_{i-1},x_{i+1}]$ such that $y[x_{i-1},x_i,x_{i+1}]$ has the value $\frac{1}{2}y''(\xi_i)$. Further, one can prove that, in the case $y_i = y(x_i)$, $i=1,2,\ldots,n$, the number of sign changes in the sequence of second differences is no greater than the number of sign changes in $y''(x)$, $x_1 \le x \le x_n$. Therefore we take the view that some smoothing of the data may be helpful if their second differences have more sign changes than are expected in the second derivative of the underlying function.

The smoothing method of this paper seeks real numbers $\phi_i$, $i=1,2,\ldots,n$, such that the sequence

$$\phi[x_{i-1},x_i,x_{i+1}] = \frac{\phi_{i-1}}{(x_{i-1}-x_i)\,(x_{i-1}-x_{i+1})} + \frac{\phi_i}{(x_i-x_{i-1})\,(x_i-x_{i+1})} + \frac{\phi_{i+1}}{(x_{i+1}-x_{i-1})\,(x_{i+1}-x_i)}, \qquad i=2,3,\ldots,n-1, \tag{1.2}$$

changes sign at most $k-1$ times, where $k$ is a positive integer that is also provided by the user. Ideally, $k-1$ is the number of sign changes in the second derivative of the underlying function that has just been mentioned, but of course the user can try many values of $k$ that may be suitable. The calculated numbers $\phi_i$, $i=1,2,\ldots,n$, are smoothed values of the input $y_i$, $i=1,2,\ldots,n$. Specifically, the method minimizes the sum of squares

$$F(\underline{\phi}) = \textstyle\sum_{i=1}^{n} (y_i-\phi_i)^2, \qquad \underline{\phi}\in\mathcal{R}^n, \tag{1.3}$$

subject to the constraints on the components of $\underline{\phi}$ that have been stated. This calculation is proposed and studied in the dissertation of Demetriou (1985). Moreover, the minimization of the objective function $\max\{|y_i-\phi_i| : i=1,2,\ldots,n\}$, subject to the same constraints on $\underline{\phi}\in\mathcal{R}^n$, is addressed by Cullinan (1986).

Let the points $(x_i, \phi_i)$, $i=1,2,\ldots,n$, be plotted on a piece of paper, and, for $i=1,2,\ldots,n-1$, let $(x_i, \phi_i)$ be joined to $(x_{i+1}, \phi_{i+1})$ by a straight line. We introduce the following definition of a 'convex section' of this piecewise linear curve. Let $p$ and $q$ be any integers from the sets $\{1,3,4,\ldots,n-1\}$ and $\{2,3,\ldots,n-2,n\}$, respectively, such that $p \leq q$. We say that the part of the piecewise linear curve on the closed interval $[x_p, x_q]$ is a 'convex section' if the following three conditions are satisfied.

(C1) The second differences $\phi[x_{i-1}, x_i, x_{i+1}]$, $i \in [p,q] \cap [2, n-1]$, are nonnegative and at least one of them is positive.

(C2) Either $p=1$, or there is at least one nonzero number in the sequence $\phi[x_{i-1}, x_i, x_{i+1}]$, $i=2,3,\ldots,p-1$, and the last of them is negative.

(C3) Either $q=n$, or there is at least one nonzero number in the sequence $\phi[x_{i-1}, x_i, x_{i+1}]$, $i=q+1, q+2, \ldots, n-1$, and the first of them is negative.

We define a 'concave section' similarly, replacing 'nonnegative', 'positive' and 'negative' by 'nonpositive', 'negative' and 'positive', respectively, in the last three conditions. It follows that, if the sequence of differences (1.2) changes sign $k-1$ times, then there exist $k$ disjoint intervals $[x_{p(j)}, x_{q(j)}]$, $j=1,2,\ldots,k$, such that each of the data values $x_i$, $i=1,2,\ldots,n$, is in exactly one of them, and such that each interval provides either a 'convex section' or a 'concave section' of the piecewise linear curve that has been mentioned.

The definitions of the last paragraph allow a highly useful property of the required fit to be stated. Indeed, if $\underline{\phi} \in \mathcal{R}^n$ is optimal, and if this fit has a 'convex section' on $[x_p, x_q]$, then the components $\phi_i$, $i=p, p+1, \ldots, q$, are defined by the data $(x_i, y_i)$, $i=p, p+1, \ldots, q$. Specifically, these components have the values that minimize the sum of squares $\sum_{i=p}^{q} (y_i - \phi_i)^2$ subject to the constraints $\phi[x_{i-1}, x_i, x_{i+1}] \geq 0$, $i=p+1, p+2, \ldots, q-1$, except that there are no constraints in the case $q \leq p+1$. Similarly, if an optimal $\underline{\phi}$ has a 'concave section' on $[x_p, x_q]$, then the components $\phi_i$, $i=p, p+1, \ldots, q$, minimize $\sum_{i=p}^{q} (y_i - \phi_i)^2$, subject to the concavity of the piecewise linear curve through the points $(x_i, \phi_i)$, $i=p, p+1, \ldots, q$. Therefore, if one knows (or guesses) the convex and concave sections of an optimal fit, then the components of $\underline{\phi}$ are defined by a separate quadratic programming problem for each section. We see that, if the section is $[x_p, x_q]$, then the quadratic programming problem has only $q-p+1$ variables. Further, its second derivative matrix is twice the unit matrix, and the matrix of coefficients of its linear constraints is part of a tridiagonal matrix. An algorithm for its solution, that takes advantage of all this structure, is proposed by Demetriou and Powell (1991). Thus one can calculate $\alpha[p,q]$ and $\beta[p,q]$ very efficiently, where these numbers are defined to be the least value of $\sum_{i=p}^{q} (y_i - \phi_i)^2$, subject to the piecewise linear curve through the points $(x_i, \phi_i)$, $i=p, p+1, \ldots, q$, being convex or concave, respectively.

For example, we consider the minimization of the sum of squares (1.3) when one sign change is allowed in the second differences (1.2), and when any positive second differences have to precede any negative ones. Letting $\underline{\phi} \in \mathcal{R}^n$ be optimal, we assume that both positive and negative second differences occur, and we let $q$ be an integer in $[2, n-2]$ such that there is a 'convex section' of $\underline{\phi}$ on the interval $[x_1, x_q]$, which implies that there is a 'concave section' on $[x_{q+1}, x_n]$. It follows from the previous paragraph that the optimal value of the objective function is the sum

$$F(\underline{\phi}) = \alpha[1, q] + \beta[q+1, n]. \tag{1.4}$$

Further, because $q$ is not known in advance, one can calculate the right hand side of this sum for every $q$ in $[2, n-2]$ in order to find an optimal $q$. This technique is recommended, but it is important to note that the constraints may exclude some of the trial values of $q$. Indeed, if $n=6$ and if $\underline{y}$ has the components $y_i = (x_i - x_1)^2$, $i = 1, 2, 3$, and $y_i = -(x_6 - x_i)^2$, $i = 4, 5, 6$, then $\alpha[1, 3] + \beta[4, 6]$ is zero, but the choice $\underline{\phi} = \underline{y}$ is excluded by the fact that the differences $y[x_1, x_2, x_3]$, $y[x_2, x_3, x_4]$, $y[x_3, x_4, x_5]$ and $y[x_4, x_5, x_6]$ alternate in sign.

The assertion that the components of $\underline{\phi}$ can be generated by solving a separate quadratic programming problem on each convex and concave section is proved in Section 2. This important property is taken from Chapter 6 of Demetriou (1985), except that the results of Section 2 are stronger in the case when some zero second differences occur between nonzero ones at the sign changes of the sequence (1.2). The theory implies that the required fit can be generated efficiently by a dynamic programming method, that extends the use of equation (1.4) to general values of $k$. Some algorithms of this kind are proposed by Demetriou (1985). They require only $\mathcal{O}(kn^2)$ computer operations after all the relevant values of $\alpha[p, q]$ and $\beta[p, q]$, $1 \le p \le q \le n$, have been calculated. Section 3 considers the main techniques that are employed, and also one of these algorithms is described in detail because it is useful to Section 5.

The success of the dynamic programming method of Section 3 depends on a conjecture that hardly ever fails. The conjecture includes the claim that, if $\underline{\phi} \in \mathcal{R}^n$ is the best fit to the data, and if this fit has a 'convex section' and a 'concave section' on $[x_p, x_q]$ and on $[x_{q+1}, x_r]$, respectively, then $\phi_i$, $i = p, p+1, \ldots, r$, is the best fit to the data $y_i$, $i = p, p+1, \ldots, r$, subject to any positive second differences of the fit preceding any negative ones. The claim is plausible, because the components $\phi_i$, $i = p, p+1, \ldots, r$, satisfy these constraints, and they have been chosen to minimize $\sum_{i=p}^{r} (y_i - \phi_i)^2$, except that any other convex and concave sections of $\underline{\phi}$ may cause some interference, although the analysis of Section 2 establishes that single convex and concave sections do not interact. Further, all the relevant numerical experiments of the authors during the last 12 years have corroborated the conjecture. Recently, however, we tried to prove the conjecture when only two sign changes are allowed in the second differences (1.2). Unfortunately, this work yielded a counter-example, which is the subject of Section 4. If one of the algorithms of Demetriou (1985) were applied to the counter-example, then the resultant fit would satisfy the divided difference constraints, but the sum of squares of residuals would be about 0.2% greater than the least value that can be achieved.

Therefore an extension to the dynamic programming methods of Section 3 is proposed in Section 5. By employing more computer storage and more computation, it provides a fit $\phi_i$, $i = 1,2,\ldots,n$, to the data $(x_i, y_i)$, $i = 1,2,\ldots,n$, that is guaranteed to minimize the sum of squares (1.3), subject to the given constraints on the signs of the second differences (1.2).

# 2 Separation properties of convex and concave sections

Throughout this section we let $\phi_i$, $i=1,2,\ldots,n$, be a best fit to the data $y_i$, $i=1,2,\ldots,n$, and we let $k-1$ be the number of sign changes in the sequence (1.2) of second divided differences, where $k \geq 2$. We assume without loss of generality that the sign of the first nonzero second difference of the fit is positive. We have found already that we can let $p(j)$ and $q(j)$, $j=1,2,\ldots,k$, be integers such that $p(j) \leq q(j)$, $j=1,2,\ldots,k$, $p(1)=1$, $p(j+1)=q(j)+1$, $j=1,2,\ldots,k-1$, and $q(k)=n$, and such that there is a 'convex section' or a 'concave section' of the fit on $[x_{p(j)}, x_{q(j)}]$ when $j$ is odd or even, respectively. The choice of $p(j)$, $j=2,3,\ldots,k$, is not unique if some zero second differences occur between nonzero ones at the sign changes of the sequence (1.2), and we require a notation that can show this nonuniqueness. Therefore, for $j = 1,2,\ldots,k$, we let $\hat{p}(j)$ and $\check{p}(j)$ be the least and greatest values of $p(j)$ that are allowed by the conditions on $p(j)$ that have just been stated. Similarly, for each $j$, we let $\hat{q}(j)$ and $\check{q}(j)$ be the least and greatest values of $q(j)$ that are allowed.

An equivalent definition of these integers is as follows. We set the values $\hat{p}(1)=\check{p}(1)=1$ and $\hat{q}(k)=\check{q}(k)=n$. Moreover, for $j=1,2,\ldots,k-1$, we let $\check{q}(j)$ be the greatest integer such that there are only $j-1$ sign changes in the sequence $\phi[x_{i-1}, x_i, x_{i+1}]$, $i=2,3,\ldots,\check{q}(j)$. Furthermore, $\hat{q}(j)$ is the greatest integer that satisfies $\hat{q}(j) \leq \check{q}(j)$ and the condition

$$\phi[x_{\hat{q}(j)-1}, x_{\hat{q}(j)}, x_{\hat{q}(j)+1}] \neq 0, \qquad j=1,2,\ldots,k-1. \tag{2.1}$$

Then, for $j=2,3,\ldots,k$, the integer $\hat{p}(j)$ is as small as possible subject to no sign changes in the sequence $\phi[x_{i-1}, x_i, x_{i+1}]$, $i=\hat{p}(j), \hat{p}(j)+1, \ldots, \check{q}(j)$, and $\check{p}(j)$ is the least integer that satisfies $\check{p}(j) \geq \hat{p}(j)$ and the condition

$$\phi[x_{\check{p}(j)-1}, x_{\check{p}(j)}, x_{\check{p}(j)+1}] \neq 0, \qquad j=2,3,\ldots,k. \tag{2.2}$$

It follows that, for any integer $j$ in $[1,k]$, the range $[\check{p}(j),\widehat{q}(j)]$ is the shortest of the intervals $[p(j),q(j)]$ that can occur in the construction of the previous paragraph.

Using this notation, the important property of $\underline{\phi}\in\mathcal{R}^n$ that is stated in Section 1 and that will be proved in this section is given in Theorem 2.1 below. Our method of analysis will also establish Theorem 2.2. The proofs of these two theorems will occupy the remainder of this section.

**Theorem 2.1** *We assume the conditions and we employ the notation of the first two paragraphs of this section. Let $j$ be any integer from $[1,k]$, and let $p$ and $q$ be any integers from the intervals $[\widehat{p}(j),\check{p}(j)]$ and $[\widehat{q}(j),\check{q}(j)]$, respectively. Then the components $\phi_i$, $i=p,p+1,\ldots,q$, have the values that solve the quadratic programming problem:*

$$\left.\begin{array}{lll} \textit{Minimize} & \sum_{i=p}^{q}(y_i-\phi_i)^2, & \\ \textit{subject to} & (-1)^j\,\phi[x_{i-1},x_i,x_{i+1}]\le 0, & i=p+1,p+2,\ldots,q-1, \end{array}\right\}\qquad(2.3)$$

*except that there are no constraints if $q\le p+1$.*

**Theorem 2.2** *If $j$ is any integer from $[1,k-1]$ such that $\widehat{q}(j)$ is strictly less than $\check{q}(j)$, then the interpolation equations $\phi_i=y_i$, $i=\widehat{q}(j)+1,\widehat{q}(j)+2,\ldots,\check{q}(j)$, are satisfied.*

The proof of Theorem 2.1 is simple in the case $\widehat{p}(j)=\check{p}(j)=\widehat{q}(j)=\check{q}(j)=p$, say, because then the differences $\phi[x_{p-2},x_{p-1},x_p]$, $\phi[x_{p-1},x_p,x_{p+1}]$ and $\phi[x_p,x_{p+1},x_{p+2}]$ are all nonzero and they alternate in sign. Thus there exists $\varepsilon>0$ such that, if $\phi_p$ is perturbed by any number whose modulus is less than $\varepsilon$, and if the other components of $\underline{\phi}$ are unaltered, then the signs of all the differences $\phi[x_{i-1},x_i,x_{i+1}]$, $i=2,3,\ldots,n-1$, are preserved. Therefore such a perturbation to $\underline{\phi}$ is not allowed to reduce the sum of squares (1.3). It follows that $\phi_p=y_p$ is achieved, which is the statement of Theorem 2.1 when $p=q$. This argument shows some of the features of the proof of the following lemma, which is taken from Demetriou (1985).

**Lemma 2.3** *Theorem 2.1 is true when $p=\check{p}(j)$ and $q=\widehat{q}(j)$.*

**Proof** We assume without loss of generality that the integer $j$ is odd. It follows from the first paragraph of this section that the inequalities

$$\phi[x_{i-1},x_i,x_{i+1}]\ge 0,\qquad i\in[p,q]\cap[2,n-1],\qquad(2.4)$$

are satisfied. Moreover, expressions (2.2) and (2.1) provide the conditions

$$\left.\begin{array}{lll} p=1 & \text{or} & \phi[x_{p-1},x_p,x_{p+1}]>0, \\ q=n & \text{or} & \phi[x_{q-1},x_q,x_{q+1}]>0. \end{array}\right\}\qquad(2.5)$$

We suppose that Lemma 2.3 is false, so there exist numbers $\psi_i$, $i = p$, $p+1, \ldots, q$, such that the piecewise linear curve through the points $(x_i, \psi_i)$, $i=p, p+1, \ldots, q$, is convex, and such that the strict inequality

$$\textstyle\sum_{i=p}^{q} (y_i - \theta\psi_i - \{1-\theta\}\phi_i)^2 < \sum_{i=p}^{q} (y_i - \phi_i)^2 \qquad (2.6)$$

is satisfied when $\theta=1$. The reason for introducing the parameter $\theta$ is that we consider replacing $\phi_i$ by $(1-\theta)\phi_i + \theta\psi_i$ for $i=p, p+1, \ldots, q$, but we do not disturb the other components of $\underline{\phi} \in \mathcal{R}^n$. We preserve the conditions (2.5) by choosing $\theta$ to be a sufficiently small positive number. It follows from the constraints $\psi[x_{i-1}, x_i, x_{i+1}] \geq 0$, $i=p+1, p+2, \ldots, q-1$, when $q \geq p+2$, and from $\theta \in [0,1]$, that the change to $\underline{\phi} \in \mathcal{R}^n$ also preserves the conditions (2.4). Moreover, because the left hand side of expression (2.6) is a convex function of $\theta$ that is equal to the right hand side when $\theta=0$, we infer that the alteration of $\underline{\phi}$ provides a strict reduction in the objective function $\sum_{i=1}^{n}(y_i-\phi_i)^2$. Therefore we have found a contradiction that proves the lemma, provided that the alteration does not increase the number of sign changes in the sequence (1.2).

Our perturbation to $\underline{\phi}$ preserves not only the inequalities (2.4) and (2.5), but also the value of $\phi[x_{i-1}, x_i, x_{i+1}]$ when $i \leq p-2$ or $i \geq q+2$, because, for these integers $i$, the relevant components of $\underline{\phi}$ have not been altered. Therefore the contradiction can be prevented only by the sign of $\phi[x_{p-2}, x_{p-1}, x_p]$ when $p \geq 3$ or by the sign of $\phi[x_q, x_{q+1}, x_{q+2}]$ when $q \leq n-2$. In the first case, the original $\underline{\phi}$ has a 'concave section' just before the interval $[x_p, x_q]$. Therefore, if we look backwards through the original sequence of second differences starting with $\phi[x_{p-2}, x_{p-1}, x_p]$, then the first nonzero value that is encountered has a negative sign. It follows, from the strict positivity of $\phi[x_{p-1}, x_p, x_{p+1}]$ when $j$ is odd and $p \geq 3$, that the change to $\phi[x_{p-2}, x_{p-1}, x_p]$ does not increase the number of sign changes in the sequence (1.2). A similar argument can be applied to $\phi[x_q, x_{q+1}, x_{q+2}]$ when $q \leq n-2$. Therefore the lemma is true. □

Next we address the proof of Theorem 2.2. A technique that will be used is easy to explain when $\widehat{q}(j) = \check{q}(j)-1$. In this case, under the usual assumption that $j$ is odd, expression (2.1) and the definition of $\check{q}(j)$ provide the strict inequalities

$$\phi[x_{\check{q}(j)-2}, x_{\check{q}(j)-1}, x_{\check{q}(j)}] > 0 \quad \text{and} \quad \phi[x_{\check{q}(j)}, x_{\check{q}(j)+1}, x_{\check{q}(j)+2}] < 0, \qquad (2.7)$$

respectively. Further, there exists $\varepsilon > 0$ such that, if $\phi_{\check{q}(j)}$ is perturbed by any amount whose modulus is less than $\varepsilon$, and if the other components of $\underline{\phi} \in \mathcal{R}^n$ are not changed, then the inequalities (2.7) are preserved. The only other second difference of $\underline{\phi}$ that is altered is $\phi[x_{\check{q}(j)-1}, x_{\check{q}(j)}, x_{\check{q}(j)+1}]$, but expression (2.7) shows that its value is irrelevant to the number of sign changes in the sequence (1.2). Therefore, because $\underline{\phi}$ is optimal, the perturbation to $\underline{\phi}$ does not increase the sum of squares $\sum_{i=1}^{n}(y_i-\phi_i)^2$. It follows that the interpolation equation $\phi_{\check{q}(j)} = y_{\check{q}(j)}$ is satisfied as required.

**Proof of Theorem 2.2.** We continue to assume without loss of generality that $j$ is an odd integer. Therefore the definitions of $\widehat{q}(j)$ and $\check{q}(j)$ imply that the optimal $\underline{\phi}\in\mathcal{R}^n$ has the properties

$$\left.\begin{array}{l} \phi[x_{\widehat{q}(j)-1},x_{\widehat{q}(j)},x_{\widehat{q}(j)+1}]>0, \quad \phi[x_{\check{q}(j)},x_{\check{q}(j)+1},x_{\check{q}(j)+2}]<0 \\ \text{and} \quad \phi[x_{i-1},x_i,x_{i+1}]=0, \quad i=\widehat{q}(j)+1,\widehat{q}(j)+2,\ldots,\check{q}(j). \end{array}\right\} \tag{2.8}$$

Let $\ell$ be any integer in $[\widehat{q}(j)+1,\check{q}(j)]$ and let $\underline{\psi}^{(\ell)}$ be the vector in $\mathcal{R}^n$ whose components are all zero except for the values

$$\left.\begin{array}{ll} \psi_i^{(\ell)}=(x_i-x_{\widehat{q}(j)})\,/\,(x_\ell-x_{\widehat{q}(j)}), & i=\widehat{q}(j)+1,\widehat{q}(j)+2,\ldots,\ell, \\ \psi_i^{(\ell)}=(x_{\check{q}(j)+1}-x_i)\,/\,(x_{\check{q}(j)+1}-x_\ell), & i=\ell,\ell+1,\ldots,\check{q}(j). \end{array}\right\} \tag{2.9}$$

Further, let $\varepsilon$ be a positive constant such that, if we replace $\underline{\phi}$ by $\underline{\phi}+\theta\underline{\psi}^{(\ell)}$, where $\theta$ is any real number that satisfies $|\theta|<\varepsilon$, then the first line of expression (2.8) is preserved. Now $\underline{\psi}^{(\ell)}$ has been chosen so that the difference $\psi^{(\ell)}[x_{i-1},x_i,x_{i+1}]$, $i\in[2,n-1]$, is nonzero only if $i$ is one of the three integers $\widehat{q}(j)$, $\ell$ and $\check{q}(j)+1$. Thus the last line of expression (2.8) is also preserved except when $i=\ell$. Therefore it follows from the first line of this expression that, for the new choice of $\underline{\phi}$, the number of sign changes in the sequence (1.2) is independent of the value of $\phi[x_{\ell-1},x_\ell,x_{\ell+1}]$. Hence the least value of the sum of squares $\sum_{i=1}^n(y_i-\phi_i-\theta\psi_i^{(\ell)})^2$ is achieved when $\theta=0$. In other words, $\underline{\phi}$ satisfies the equations

$$(\underline{y}-\underline{\phi})^T\underline{\psi}^{(\ell)}=\sum_{i=1}^n(y_i-\phi_i)\,\psi_i^{(\ell)}=0, \qquad \ell=\widehat{q}(j)+1,\widehat{q}(j)+2,\ldots,\check{q}(j). \tag{2.10}$$

Now, by considering the second differences of $\underline{\psi}^{(\ell)}$, it is straightforward to prove that the vectors $\underline{\psi}^{(\ell)}$, $\ell=\widehat{q}(j)+1,\widehat{q}(j)+2,\ldots,\check{q}(j)$, are linearly independent. Therefore these vectors span the linear space of vectors in $\mathcal{R}^n$ whose first $\widehat{q}(j)$ and last $n-\check{q}(j)$ components are zero. In particular, the coordinate vectors $\underline{e}_i$, $i=\widehat{q}(j)+1,\widehat{q}(j)+2,\ldots,\check{q}(j)$, belong to this space. It follows from equation (2.10) that the residual vector $\underline{y}-\underline{\phi}$ is orthogonal to each of these coordinate vectors, which provides the interpolation equations that are stated in Theorem 2.2. □

Theorem 2.2 is equivalent to the statement that, if $j$ is any integer from $[2,k]$ such that $\widehat{p}(j)$ is strictly less than $\check{p}(j)$, then $\underline{\phi}$ enjoys the interpolation properties $\phi_i=y_i$, $i=\widehat{p}(j),\widehat{p}(j)+1,\ldots,\check{p}(j)-1$. A suitable way of proving this assertion is to deduce from the construction of the second paragraph of this section that, for $j\in[2,k]$, the integers $\widehat{p}(j)$ and $\check{p}(j)$ take the values $\widehat{q}(j-1)+1$ and $\check{q}(j-1)+1$, respectively.

We now turn to the proof of Theorem 2.1. In view of Lemma 2.3, it remains to consider the case when one (or both) of the conditions $p<\check{p}(j)$ and $q>\hat{q}(j)$ occurs. It is important to note that Theorem 2.2 provides the identities

$$\phi_i = y_i, \qquad i \in [p,q] \setminus [\check{p}(j), \hat{q}(j)]. \tag{2.11}$$

**Proof of Theorem 2.1.** We consider the following quadratic programming problem. Find the components $\phi_i$, $i = p, p+1, \ldots, q$, that minimize $\sum_{i=p}^{q}(y_i - \phi_i)^2$ subject to the conditions

$$(-1)^j \, \phi[x_{i-1}, x_i, x_{i+1}] \leq 0, \qquad i = \check{p}(j)+1, \check{p}(j)+2, \ldots, \hat{q}(j)-1. \tag{2.12}$$

We see that there are no constraints on $\phi_i$ if $i$ is any integer from $[p,q]$ that is not in the interval $[\check{p}(j), \hat{q}(j)]$. Therefore in this case $\phi_i$ has the value (2.11). Furthermore, we see that the required components $\phi_i$, $i=\check{p}(j), \check{p}(j)+1, \ldots, \hat{q}(j)$, are given by the quadratic programming problem of Theorem 2.1 when $p = \check{p}(j)$ and $q = \hat{q}(j)$. It follows from Theorem 2.2 and Lemma 2.3 that the solution to the quadratic programming problem of this paragraph is achieved by the components $\phi_i$, $i=p, p+1, \ldots, q$, of the vector $\underline{\phi} \in \mathcal{R}^n$ that solves the data fitting calculation of Section 1.

It remains to show that these components are also the solution to the quadratic programming problem (2.3). They satisfy the conditions in the last line of expression (2.3), because $p$ and $q$ are chosen so that $[x_p, x_q]$ provides a 'convex section' or 'concave section' of $\underline{\phi} \in \mathcal{R}^n$ when $j$ is odd or even, respectively. Furthermore, because the constraints (2.12) are a subset of the constraints (2.3), the optimal value of $\sum_{i=p}^{q}(y_i - \phi_i)^2$ in the quadratic programming problem of the previous paragraph is a lower bound on the optimal value of this sum of squares in the problem (2.3). Therefore the components $\phi_i$, $i=p, p+1, \ldots, q$, solve the problem (2.3) as required, which completes the proof. □

# 3 A dynamic programming algorithm

Most of the techniques of the algorithm of this section are taken from Demetriou (1985). The user has to provide the data $(x_i, y_i)$, $i=1,2,\ldots,n$, and the integer $k$, that are introduced in the first two paragraphs of Section 1, where $k$ satisfies $k \leq n-1$. Furthermore, a parameter, $s$ say, has to be set to $+1$ or $-1$. Then the vector $\underline{\phi} \in \mathcal{R}^n$ is defined to be 'feasible' if one of the following three conditions is satisfied: (F1) All of the second differences (1.2) are zero, or (F2) The first nonzero element of the sequence (1.2) has the same sign as $s$ and the number of sign changes in the sequence is at most $k-1$, or (F3) The sign of the first nonzero element of the sequence (1.2) is opposite to the sign of $s$ and the number of sign changes in the sequence is at most $k$–2. The given algorithm should calculate the least value of $\sum_{i=1}^{n}(y_i-\phi_i)^2$, subject to the feasibility of $\underline{\phi}$. We assume that $s=+1$ is set, so it is usual for the first section of the piecewise linear curve through the points $(x_i,\phi_i)$, $i=1,2,\ldots,n$, to be convex. On the other hand, if the second differences $y[x_{i-1},x_i,x_{i+1}]$, $i=2,3,\ldots,n-1$, are all nonpositive, then the feasibility condition (F3) or (F1) allows $\underline{\phi}=\underline{y}$ when $k\geq 2$.

The given algorithm depends on the separation properties of convex and concave sections that are proved in Section 2. Therefore it calculates integers $p(j)$ and $q(j)$, $j=1,2,\ldots,k$, such that usually the final fit has a 'convex section' on $[x_{p(j)},x_{q(j)}]$ when $j$ is odd and a 'concave section' on $[x_{p(j)},x_{q(j)}]$ when $j$ is even. Here the word 'usually' allows for the possibility that $\underline{\phi}$ may satisfy the feasibility condition (F3) that is given in the previous paragraph, and it covers some other situations that are explained in Section 5. Indeed, provided that we restrict attention to trial approximations that satisfy the constraints, and provided that we do not exclude the required solution, we can bend the rules for $p$ and $q$ that are given in Section 1. In particular, in the case of the feasibility condition (F3), the equation $\phi_1=y_1$ must hold, so the algorithm sets the values $p(1)=q(1)=1$ and $p(2)=2$. Thus the fit can begin with a concave section on the interval $[x_1,x_{q(2)}]$. Further, as indicated in the first paragraph of Section 2, the calculated integers always have the properties

$$\left.\begin{array}{l} p(1)=1, \quad q(k)=n, \quad p(j)\leq q(j), \quad j=1,2,\ldots,k, \\ \text{and} \qquad p(j+1)=q(j)+1, \quad j=1,2,\ldots,k-1. \end{array}\right\} \tag{3.1}$$

After the algorithm has decided on their values, then all of the components of $\underline{\phi}\in\mathcal{R}^n$ are defined by solving the quadratic programming problem (2.3) with $p=p(j)$ and $q=q(j)$ for $j=1,2,\ldots,k$.

We recall from Section 1 the notation $\alpha[p,q]$ and $\beta[p,q]$ for the least value of the objective function of the problem (2.3) when $j$ is odd or even, respectively. It follows from the previous paragraph that the sum of squares of residuals of the calculated fit has the value

$$\begin{aligned}\sum_{i=1}^{n}(y_i-\phi_i)^2 &= \alpha[p(1),q(1)]+\beta[p(2),q(2)]+\alpha[p(3),q(3)]+\cdots \\ &\qquad \cdots+(\alpha,\beta;k)[p(k),q(k)], \end{aligned}\tag{3.2}$$

where $(\alpha,\beta;k)$ denotes $\alpha$ when $k$ is odd and $\beta$ when $k$ is even. The techniques of this section also make use of the remark that expression (3.1) implies $q(j)\geq j$, $j=1,2,\ldots,k$.

A method for calculating a best fit when $k=2$ is suggested in the paragraph that includes equation (1.4). Indeed, guided by formula (3.2), we seek an integer $q(1)$ that minimizes the sum

$$\alpha[1,q(1)]+\beta[p(2),n], \qquad q(1)\in\mathcal{F}[n;2], \tag{3.3}$$

where $p(2)=q(1)+1$ and where the elements of $\mathcal{F}[n;2]$ satisfy a condition for feasibility. We always put the integer $n-1$ in $\mathcal{F}[n;2]$, in order that expression (3.3) can be the sum of squares of residuals of the best convex fit to the data whenever that fit is relevant. Moreover, we put $q(1)\in\{1,2,\ldots,n-2\}$ in $\mathcal{F}[n;2]$ if the following test is satisfied. Let the components $\phi_i$, $i=1,2,\ldots,q(1)$, solve the quadratic programming problem (2.3) when $j=1$, $p=1$ and $q=q(1)$, and let the remaining components of $\underline{\phi}$ solve that problem when $j=2$, $p=p(2)$ and $q=n$. Thus $\alpha[1,q(1)]+\beta[p(2),n]$ is the sum $\sum_{i=1}^{n}(y_i-\phi_i)^2$. Then we include $q(1)\in[1,n-2]$ in $\mathcal{F}[n;2]$ if and only if the inequality $\phi[x_{q(1)},x_{p(2)},x_{p(2)+1}]\leq 0$ is achieved, which ensures feasibility, and which can capture the optimal value of $q(1)$, because the required fit has a 'concave section' on $[x_{p(2)},x_n]$ in the case $q(1)\leq n-2$. Therefore the fit $\phi_i$, $i=1,2,\ldots,n$, is optimal when $q(1)$ minimizes the sum (3.3). The fact that this test for feasibility does not depend on the sign of $\phi[x_{q(1)-1},x_{q(1)},x_{p(2)}]$ will be useful later.

A crucial feature of this technique is that, for each trial value of $q(1)$ in $[1,n-1]$, the first $q(1)$ components of $\underline{\phi}$ depend only on the data $(x_i,y_i)$, $i=1,2,\ldots,q(1)$, and the other components of $\underline{\phi}$ are defined by the data $(x_i,y_i)$, $i=p(2),p(2)+1,\ldots,n$. Our algorithm is based on the expectation that this feature extends to $k\geq 3$ in the following way. Let $j$ be any integer from $[1,k]$, and let $q(j)$ have the value that occurs in the optimal sum (3.2). Then it is assumed that the first $q(j)$ components of $\underline{\phi}\in\mathcal{R}^n$ can be derived just from the data $(x_i,y_i)$, $i=1,2,\ldots,q(j)$, and from the value of $j$. The assumption is trivial when $j=k$, because in this case we have $q(j)=n$. Furthermore, Theorem 2.1 establishes that it is true when $j=1$. If it failed for an intermediate value of $j$, however, then our algorithm would achieve the required constraints on the signs of the second differences (1.2), but the sum of squares of residuals of the final fit would not be optimal. This unusual deficiency is demonstrated by the counter-example of Section 4.

The assumption makes the following procedure suitable when $k=3$. We are supposing that the components $\phi_i$, $i=1,2,\ldots,q(2)$, can be calculated from the data $(x_i,y_i)$, $i=1,2,\ldots,q(2)$, so they must have the values that minimize $\sum_{i=1}^{q(2)}(y_i-\phi_i)^2$, subject to the condition that any positive second differences in the sequence $\phi[x_{i-1},x_i,x_{i+1}]$, $i=2,3,\ldots,q(2)-1$, come before any negative ones. Therefore we regard $q(2)$ as the value of $n$ for the moment, in order that the first $q(2)$ components of $\underline{\phi}$ can be generated by the method of the paragraph that includes equation (3.3). Then the last $n-q(2)$ components of $\underline{\phi}$ are calculated as usual by solving the appropriate quadratic programming problem (2.3). Thus the feasibility of $\underline{\phi}$ should be achieved automatically when $q(2)$ is optimal. Now this value of $q(2)$ is in the set $\{2,3,\ldots,n-1\}$, so we let $q(2)$ run through these integers, and for each trial $q(2)$ we apply the technique that has just been described. If $q(2)\leq n-2$, and if the second difference $\phi[x_{q(2)},x_{q(2)+1},x_{q(2)+2}]$ of the resultant $\underline{\phi}$ is negative, then, according to the theory of Section 2, the current $q(2)$ is not the integer that we require.

In all other cases, however, we note the sum of squares $\sum_{i=1}^{n}(y_i - \phi_i)^2$, and we pick a value of $q(2)$ that minimizes the sum, because, if the assumption is true, then $\underline{\phi}\in\mathcal{R}^n$ is the best fit to the data.

We extend this procedure to larger values of $k$ by recursion. Indeed, if we know how to calculate a best fit when $k$ is replaced by $k-1$, then we seek the optimal value of $q(k-1)$ in the interval $[k-1, n-1]$. Each trial value of this integer is regarded as the value of $n$ for the moment, and the known procedure is applied to the data $(x_i, y_i)$, $i=1,2,\ldots,q(k-1)$, which gives the first $q(k-1)$ components of a trial fit $\underline{\phi}\in\mathcal{R}^n$. Further, the other components of $\underline{\phi}$ are defined by the quadratic programming problem (2.3) when $p=q(k-1)+1$, $q=n$ and $j=k$. We pick the value of $q(k-1)$ that minimizes $\sum_{i=1}^{n}(y_i - \phi_i)^2$ subject to the feasibility constraint

$$(-1)^k\,\phi[x_{q(k-1)}, x_{q(k-1)+1}, x_{q(k-1)+2}] \leq 0 \quad \text{or} \quad q(k-1)=n-1. \tag{3.4}$$

The implementation of this procedure by Demetriou (1985) is efficient, because a dynamic programming method can be employed. In order to describe it, we let $j$ be any integer from $[1, k-1]$, we let $\bar{q}$ be any trial value of $q(j)$ in equation (3.2), we suppose that the components $\phi_i$, $i=1,2,\ldots,\bar{q}$, are generated by the recursive procedure that has been specified already for deducing the first $j$ terms on the right hand side of expression (3.2), and we define $\gamma[\bar{q};j]=\sum_{i=1}^{\bar{q}}(y_i-\phi_i)^2$ to be the resultant value of this partial sum. Therefore the formula

$$\gamma[\bar{q};1] = \alpha[1,\bar{q}], \qquad \bar{q}\in\{1,2,\ldots,n\}, \tag{3.5}$$

provides the starting values for the method. Further, it follows from the paragraph that includes equation (3.3), that we can also employ the formula

$$\gamma[\bar{q};2] = \min\{\gamma[q;1] + \beta[p,\bar{q}] : q\in\mathcal{F}[\bar{q};2]\}, \qquad \bar{q}\in\{2,3,\ldots,n\}, \tag{3.6}$$

where $p=q+1$ and where $\mathcal{F}[\bar{q};2]$ is a subset of $\{1,2,\ldots,\bar{q}-1\}$. Specifically, it is the subset that is defined after equation (3.3) if we replace $q(1)$, $p(2)$ and $n$ by $q$, $p$ and $\bar{q}$, respectively.

Now the definition of $\gamma[\bar{q};j]$ in terms of the recursive procedure implies that, for all $j\geq 2$, the relation (3.6) has a generalization of the form

$$\gamma[\bar{q};j] = \min\{\gamma[q;j-1] + (\alpha,\beta;j)[p,\bar{q}] : q\in\mathcal{F}[\bar{q};j]\}, \quad \bar{q}\in\{j,j+1,\ldots,n\}, \tag{3.7}$$

where $(\alpha,\beta;j)$ still denotes $\alpha$ or $\beta$ when $j$ is odd or even, respectively, and where $p=q+1$. Further, $\mathcal{F}[\bar{q};j]$ is composed of the integer $\bar{q}-1$ and a subset of $\{j-1,j,\ldots,\bar{q}-2\}$. In order to provide a useful specification of this subset for $j\geq 3$, we pick the value of $\phi_q$ that occurs in the calculation of $\gamma[q;j-1]$, which we denote by $\phi_{q;j-1}$, and we take the components $\phi_p$ and $\phi_{p+1}$ from the fit that provides the sum of squares $(\alpha,\beta;j)[p,\bar{q}]$. Then the condition that is necessary and sufficient for $q\in[j-1,\bar{q}-2]$ to be in the set $\mathcal{F}[\bar{q};j]$ is the inequality $(-1)^j \phi[x_q, x_p, x_{p+1}] \le 0$, which agrees with the constraint (3.4). Therefore we note for future use the values of $\gamma[\bar{q}; j]$ and $\phi_{\bar{q};j}$ whenever the right hand side of expression (3.7) is calculated. We also store $q[\bar{q}; j]$, which is defined to be the chosen value of $q$ in the set $\mathcal{F}[\bar{q}; j]$, but we discard the other numbers that occur when formula (3.7) is applied.

It follows from equation (3.2), and from the recursive use of formula (3.7), that the sum of squares of residuals of the optimal fit to the data $(x_i, y_i)$, $i=1,2,\ldots,n$, has the value

$$\gamma[n; k] = \min\{\gamma[q; k-1] + (\alpha, \beta; k)[p, n] : q \in \mathcal{F}[n; k]\}, \qquad (3.8)$$

where we are retaining the notation of equation (3.7). Furthermore, the value of $q$ that minimizes expression (3.8), namely $q[n; k]$, is the integer $q(k-1)$ that is required in equation (3.2). Then, because $q[q(k-1); k-1]$ is the optimal value of $q$ in expression (3.7) when $j=k-1$ and $\bar{q}=q(k-1)$, it is the required value of $q(k-2)$ in the sum (3.2). Hence, using recursion, we find that the formulae

$$q(k)=n \quad \text{and} \quad q(j-1) = q[q(j); j], \quad j=k, k-1, \ldots, 2, \qquad (3.9)$$

can be applied to generate the indices of the right hand end points of all the required sections of the best fit. Finally, equation (3.1) completes the calculation of the integers of formula (3.2).

Our algorithm generates the numbers that have been mentioned, but we achieve some substantial savings in storage or computation when $k \ge 5$ by restructuring the calculation. The gain in efficiency comes from the remark that, if we seek all of the numbers (3.7) for the current $j$, then most of the terms $(\alpha, \beta; j)[p, \bar{q}]$, $j \le p \le \bar{q} \le n$, are required. Therefore, if $j$ runs through the set $\{2, 3, 4, 5\}$, for example, then these terms have to be recomputed or stored. We avoid both of these tasks, however, by adjusting $\bar{q}$ instead of $j$ in the outermost loop. Indeed, $\bar{q}$ runs through the set $\{1, 2, \ldots, n\}$, and, for each $\bar{q} \ge 3$, the numbers $\alpha[p, \bar{q}]$ and $\beta[p, \bar{q}]$, $p=1,2,\ldots,\bar{q}$, are calculated and are placed in temporary storage that can be used again when $\bar{q}$ is increased. Then we set $\gamma[\bar{q}; 1] = \alpha[1, \bar{q}]$, and we apply the formula

$$\gamma[\bar{q}; j] = \min\{\gamma[q; j-1] + (\alpha, \beta; j)[p, \bar{q}] : q \in \mathcal{F}[\bar{q}; j]\}, \quad j=2,3,\ldots,k(\bar{q}), \qquad (3.10)$$

where $p=q+1$ and where $k(\bar{q})$ is the integer $\min[\bar{q}, k-1]$ or $k$ in the case $\bar{q}<n$ or $\bar{q}=n$, respectively. This method replaces equations (3.5)–(3.8). We see that the current values of $\alpha[p, \bar{q}]$ and $\beta[p, \bar{q}]$ will become superfluous when $\bar{q}$ is increased, so the perpetual storage requirements are only $\mathcal{O}(kn)$. Indeed, it is sufficient for future applications of formula (3.10) and for equation (3.9) to retain just the numbers

$$\gamma[\bar{q}; j], \quad \phi_{\bar{q};j}, \quad j=1,2,\ldots,k(\bar{q}), \quad \text{and} \quad q[\bar{q}; j], \quad j=2,3,\ldots,k(\bar{q}), \qquad (3.11)$$

for each $\bar{q}$, where $\phi_{\bar{q};j}$ and $q[\bar{q};j]$ are defined in the paragraph that includes expression (3.7). All of the remarks of this section lead to the following algorithm for data smoothing, assuming $s=+1$.

**Algorithm 3.1**

**Step 0** The data of the algorithm are the positive integers $n$ and $k$, and the coordinates $(x_i, y_i)\in\mathcal{R}^2$, $i=1,2,\ldots,n$.

**Step 1** If $k=1$ or $k=2$, then the required fit is calculated by solving a single quadratic programming problem or by the method of the paragraph that includes expression (3.3), respectively.

**Step 2** Otherwise, when $k\geq 3$, the numbers (3.11) for $\bar{q}=1$ and $\bar{q}=2$ are given the values

$$\left.\begin{array}{l}\gamma[1;1]=\gamma[2;1]=\gamma[2;2]=0, \quad \phi_{1;1}=y_1,\\ \phi_{2;1}=\phi_{2;2}=y_2 \quad \text{and} \quad q[2;2]=1.\end{array}\right\} \qquad (3.12)$$

Then $\bar{q}$ is set to 3.

**Step 3** The quadratic programming problems that define the sums of squares $\alpha[p,\bar{q}]$ and $\beta[p,\bar{q}]$, $p=1,2,\ldots,\bar{q}$, are solved by the method of Demetriou and Powell (1991). Then the numbers (3.11) are computed and stored, using equation (3.10).

**Step 4** If $\bar{q}$ is less than $n$, it is increased by one, and then there is a branch back to Step 3.

**Step 5** Equations (3.9) and (3.1) provide the required integers $p(j)$ and $q(j)$, $j=1,2,\ldots,k$.

**Step 6** Finally, the required fit $\phi_i$, $i=1,2,\ldots,n$, is generated by solving the quadratic programming problem (2.3) for $j=1,2,\ldots,k$, the values of $p$ and $q$ being $p(j)$ and $q(j)$, respectively. □

The amount of computation of such algorithms is demonstrated by Demetriou (1985). The magnitude of the total work cannot be stated in a simple way, because the quadratic programming method of Step 3 of Algorithm 3.1 includes a starting procedure that usually provides a very good estimate of the final active set, but occasionally the number of iterations of that calculation is large. Moreover, work can be saved by taking advantage of some useful interactions between the different quadratic programming calculations of Step 3. Therefore we indicate the dependence of the complexity on $n$ by quoting some timings from the numerical experiments of Demetriou (1985). Specifically, the timings apply to a method that is similar to Algorithm 3.1, when it calculates a fit to the data

$$x_i=7(i-1)\,/\,(n-1), \quad y_i=\cos(\pi x_i)+\varepsilon_i, \qquad i=1,2,\ldots,n, \qquad (3.13)$$

| $n$ | $r=0.01$ | $r=0.1$ | $r=1$ | $r=10$ |
|---|---|---|---|---|
| 10 | 0.02 | 0.02 | 0.02 | 0.02 |
| 20 | 0.05 | 0.04 | 0.04 | 0.04 |
| 50 | 0.29 | 0.29 | 0.29 | 0.33 |
| 80 | 0.79 | 0.86 | 0.76 | 0.71 |
| 100 | 1.21 | 1.24 | 1.18 | 1.20 |
| 200 | 5.38 | 5.28 | 4.48 | 4.82 |
| 500 | 31.90 | 32.66 | 28.41 | 28.48 |

**Table 1** Some timings for the data (3.13).

where each $\varepsilon_i$ is a random number from the distribution that is uniform on $[-r,r]$, for some constant $r$. Further, the choices $s=-1$ and $k=8$ are made, in order that the constraints are consistent with the sign changes of the second derivatives of the underlying function $y(x)=\cos(\pi x)$, $0 \le x \le 7$. The experiments were run on an IBM 3081 computer, the timings in seconds for several values of $n$ and $r$ being given in Table 1. They seem to be proportional to $n^2$, approximately, which suggests that the dynamic programming method is highly suitable. Indeed, if one tried to find the optimal integers $q(j)$, $j=1,2,\ldots,7$, by considering all possible combinations in equation (3.2), and if $n$ and $r$ were large, then the number of trials would be of magnitude $n^6$.

# 4 The counter-example

The conjecture that provides the basis of the dynamic programming algorithms of Demetriou (1985) and Algorithm 3.1 includes the following assumption. Let $\underline{\phi} \in \mathcal{R}^n$ be a best fit to the data $(x_i, y_i)$, $i=1,2,\ldots,n$, that is constrained by the feasibility conditions of the first paragraph of Section 3 when $k=3$. Further, let the sequence of second differences (1.2) change sign twice, the first nonzero difference being positive, and let there be unique integers $q(1)$ and $q(2)$ such that $\underline{\phi}$ has a convex, a concave and a convex section on the intervals $[x_1, x_{q(1)}]$, $[x_{q(1)+1}, x_{q(2)}]$ and $[x_{q(2)+1}, x_n]$, respectively. Then it is supposed that the components $\phi_i$, $i=1,2,\ldots,q(2)$, minimize the sum of squares $\sum_{i=1}^{q(2)}(y_i-\phi_i)^2$, subject to the condition that any positive numbers in the sequence $\phi[x_{i-1}, x_i, x_{i+1}]$, $i=2,3,\ldots,q(2)-1$, come before any negative ones.

We will find, however, that this assumption can fail. Specifically, we consider the fit $\underline{\psi} \in \mathcal{R}^8$ to the data $(x_i, y_i)$, $i=1,2,\ldots,8$, that is shown in the first four columns of Table 2, where $\varepsilon$ is a positive constant whose value is at most 0.011. It is straightforward to verify that the numbers in the fifth

| $i$ | $x_i$ | $y_i$ | $\psi_i$ | $\delta^2\psi_i$ | $y_i-\psi_i$ |
|---|---|---|---|---|---|
| 1 | 0 | $-6$ | $-3$ | | $-3$ |
| 2 | 1 | $-6$ | $-3$ | 0 | $-3$ |
| 3 | 2 | 4 | $-3$ | 0 | 7 |
| 4 | 11 | $-4$ | $-3$ | $81-9\varepsilon$ | $-1$ |
| 5 | 12 | $6-\varepsilon$ | $6-\varepsilon$ | $-9+\varepsilon$ | 0 |
| 6 | 13 | $6-\varepsilon$ | $6-\varepsilon$ | $-1+\varepsilon$ | 0 |
| 7 | 14 | 5 | 5 | $20-20\varepsilon$ | 0 |
| 8 | 34 | 5 | 5 | | 0 |

**Table 2** The counter-example of Section 4.

column of the table are the scaled second differences

$$\begin{aligned}\delta^2\psi_i &= (x_i-x_{i-1})\,(x_{i+1}-x_{i-1})\,(x_{i+1}-x_i)\,\psi[x_{i-1},x_i,x_{i+1}]\\ &= (x_{i+1}-x_i)\,\psi_{i-1}-(x_{i+1}-x_{i-1})\,\psi_i+(x_i-x_{i-1})\,\psi_{i+1},\end{aligned} \tag{4.1}$$

$i=2,3,\ldots,n-1$, the advantage of the scaling being that many fractions are avoided in our work on the counter-example. We see that $\underline{\psi}\in\mathcal{R}^8$ does satisfy the feasibility conditions when $k=3$, and that the sum of squares of residuals has the value

$$F(\underline{\psi})=\textstyle\sum_{i=1}^{8}(y_i-\psi_i)^2=68. \tag{4.2}$$

Most of the analysis of this section addresses the optimality of $\underline{\psi}$. We begin by proving that the components $\psi_i$, $i=1,2,3,4$, solve the quadratic programming problem (2.3) when $k=p=1$ and $q=4$. Indeed, we consider the identity

$$\begin{pmatrix}6\\6\\-14\\2\end{pmatrix}=6\begin{pmatrix}1\\-2\\1\\0\end{pmatrix}+2\begin{pmatrix}0\\9\\-10\\1\end{pmatrix}, \tag{4.3}$$

because the left hand side is the gradient vector of the objective function of the quadratic programming problem when $\phi_i=\psi_i$, $i=1,2,3,4$, and the vectors on the right hand side are the gradients of the linear constraint functions $\delta^2\phi_2$ and $\delta^2\phi_3$. Thus, because the multipliers 6 and 2 on the right hand side of equation (4.3) are both positive, the Karush–Kuhn–Tucker conditions for the solution of the problem (2.3) are satisfied. It follows that, if the best fit to the data, $\underline{\phi}\in\mathcal{R}^8$ say, has a convex section on $[x_1,x_4]$, then the least value of the sum of squares of residuals is bounded below by the inequality

$$\textstyle\sum_{i=1}^{8}(y_i-\phi_i)^2\geq\sum_{i=1}^{4}(y_i-\phi_i)^2=\sum_{i=1}^{4}(y_i-\psi_i)^2=68. \tag{4.4}$$

| $i$ | $x_i$ | $y_i$ | $\phi_i$ | $\delta^2\phi_i$ | $y_i-\phi_i$ |
|---|---|---|---|---|---|
| 3 | 2 | 4 | $(1282-10\varepsilon)/466$ | | $(582+10\varepsilon)/466$ |
| 4 | 11 | $-4$ | $(1606-199\varepsilon)/466$ | 0 | $(-3470+199\varepsilon)/466$ |
| 5 | 12 | $6-\varepsilon$ | $(1642-220\varepsilon)/466$ | 0 | $(1154-246\varepsilon)/466$ |
| 6 | 13 | $6-\varepsilon$ | $(1678-241\varepsilon)/466$ | 0 | $(1118-225\varepsilon)/466$ |
| 7 | 14 | 5 | $(1714-262\varepsilon)/466$ | | $(616+262\varepsilon)/466$ |

**Table 3** The best concave fit to 5 of the middle data.

Further, $\underline{\psi}$ must be a best fit, because equation (4.2) shows that it achieves this bound. Therefore, in order to establish the optimality of $\underline{\psi}$, it is sufficient to prove that a fit that does not have a convex section on $[x_1, x_4]$ cannot be optimal.

The equations $\psi_i = y_i$, $i = 5, 6, 7, 8$, and the feasibility of $\underline{\psi} \in \mathcal{R}^8$, imply that, if $\underline{\phi} \in \mathcal{R}^8$ is optimal, then the condition

$$\textstyle\sum_{i=1}^{4} (y_i - \phi_i)^2 \leq \sum_{i=1}^{8} (y_i - \phi_i)^2 \leq \sum_{i=1}^{4} (y_i - \psi_i)^2 \tag{4.5}$$

is satisfied. Further, we deduce from the previous paragraph that, if both the second differences $\delta^2\phi_2$ and $\delta^2\phi_3$ are nonnegative, then inequality (4.5) has to hold as an equation. Hence, remembering the uniqueness of the solution of the quadratic programming problem (2.3), we find $\underline{\phi} = \underline{\psi}$. It follows that our search for an optimal fit $\underline{\phi} \in \mathcal{R}^8$ that is different from $\underline{\psi}$ is restricted to those vectors $\underline{\phi}$ that have the property that one or both of the differences $\delta^2\phi_2$ and $\delta^2\phi_3$ is negative.

Next we rule out the possibility $\delta^2\phi_2 < 0$ when $\underline{\phi}$ is optimal. If it occurred, then any sufficiently small change to $\phi_1$ of either sign would preserve feasibility, so we would have $\phi_1 = y_1 = -6$. Further, any reduction in $\phi_2$ that retained $\delta^2\phi_2 < 0$ would also preserve feasibility, because the only other change to the second differences would be a decrease in $\delta^2\phi_3$. Therefore we would also have $\phi_2 \leq y_2 = -6$. Now the conditions $\phi_1 = -6$, $\phi_2 \leq -6$ and $\delta^2\phi_2 < 0$ imply $\phi_3 < -6$. It follows from $y_3 = 4$ that the value of $\sum_{i=1}^{8} (y_i - \phi_i)^2$ would exceed 100, which contradicts optimality.

Our remarks so far show that we can restrict attention to feasible vectors $\underline{\phi} \in \mathcal{R}^8$ that satisfy $\delta^2\phi_3 < 0$. Thus $\underline{\phi}$ has the property that any negative second differences in the sequence $\delta^2\phi_i$, $i = 4, 5, 6, 7$, come before any positive ones. Hence it is sufficient to prove that all vectors $\underline{\phi}$ that have this property are afflicted with the strict condition $\sum_{i=3}^{8} (y_i - \phi_i)^2 > 68$.

Some of these vectors $\underline{\phi}$ satisfy $\delta^2\phi_i \leq 0$, $i = 4, 5, 6$. Then the sum $\sum_{i=3}^{8} (y_i - \phi_i)^2$ is bounded below by the optimal value of the objective function of the problem (2.3) when $j = 2$, $p = 3$ and $q = 7$. The solution to this problem is

| $i$ | $x_i$ | $y_i$ | $\phi_i$ | $\delta^2\phi_i$ | $y_i-\phi_i$ |
|---|---|---|---|---|---|
| 4 | 11 | $-4$ | $(5482-999\varepsilon)/1874$ | | $(-12978+999\varepsilon)/1874$ |
| 5 | 12 | $6-\varepsilon$ | $(5700-956\varepsilon)/1874$ | 0 | $(5544-918\varepsilon)/1874$ |
| 6 | 13 | $6-\varepsilon$ | $(5918-913\varepsilon)/1874$ | 0 | $(5326-961\varepsilon)/1874$ |
| 7 | 14 | 5 | $(6136-870\varepsilon)/1874$ | 0 | $(3234+870\varepsilon)/1874$ |
| 8 | 34 | 5 | $(10496-10\varepsilon)/1874$ | | $(-1126+10\varepsilon)/1874$ |

**Table 4** The best convex fit to last 5 data.

given in the fourth column of Table 3, which can be deduced from the identity

$$\begin{pmatrix} 582+10\varepsilon \\ -3470+199\varepsilon \\ 1154-246\varepsilon \\ 1118-225\varepsilon \\ 616+262\varepsilon \end{pmatrix} = (582+10\varepsilon)\begin{pmatrix} 1 \\ -10 \\ 9 \\ 0 \\ 0 \end{pmatrix} + (2350+299\varepsilon)\begin{pmatrix} 0 \\ 1 \\ -2 \\ 1 \\ 0 \end{pmatrix} + (616+262\varepsilon)\begin{pmatrix} 0 \\ 0 \\ 1 \\ -2 \\ 1 \end{pmatrix}, \qquad (4.6)$$

by analogy with equation (4.3). Further, the last column of the table provides the sum of squares of residuals

$$\textstyle\sum_{i=3}^{7} (y_i-\phi_i)^2 = (32920 - 4544\,\varepsilon + 471\,\varepsilon^2)\,/\,466 = \sigma_1(\varepsilon), \qquad (4.7)$$

say, which takes the value $\sigma_1(0) = 32920/466 \approx 70.64$. The definition (4.7) also gives the inequality $\sigma_1(\varepsilon) > 70.5$, $0 \le \varepsilon \le 0.011$. Therefore, from now on we can restrict attention to vectors $\underline{\phi}$ such that at least one of the differences $\delta^2\phi_4$, $\delta^2\phi_5$ and $\delta^2\phi_6$ is positive. We treat the remaining cases in two different ways, that are distinguished by the conditions $\delta^2\phi_5 \ge 0$ and $\delta^2\phi_5 < 0$.

In the former case that has just been mentioned, we have to study vectors $\underline{\phi}$ that satisfy $\delta^2\phi_i \ge 0$, $i=5,6,7$. Therefore $\sum_{i=3}^{8}(y_i-\phi_i)^2$ is bounded below by the optimal value of the objective function of the problem (2.3) when $j=3$, $p=4$ and $q=8$, whose solution is shown in the fourth column of Table 4, the analogue of equations (4.3) and (4.6) being the identity

$$\begin{pmatrix} 12978-999\varepsilon \\ -5544+918\varepsilon \\ -5326+961\varepsilon \\ -3234-870\varepsilon \\ 1126-10\varepsilon \end{pmatrix} = (12978-999\varepsilon)\begin{pmatrix} 1 \\ -2 \\ 1 \\ 0 \\ 0 \end{pmatrix} + (20412-1080\varepsilon)\begin{pmatrix} 0 \\ 1 \\ -2 \\ 1 \\ 0 \end{pmatrix}$$

$$+(1126-10\varepsilon)\begin{pmatrix}0\\0\\20\\-21\\1\end{pmatrix}. \qquad (4.8)$$

We see that the residuals of the last column of the table provide the sum of squares

$$\textstyle\sum_{i=4}^{8}(y_i-\phi_i)^2 = (127672-21740\,\varepsilon+1879\,\varepsilon^2)\,/\,1874 = \sigma_2(\varepsilon), \qquad (4.9)$$

say. Thus we find the conditions $\sigma_2(0)=127672/1874\approx 68.13$ and $\sigma_2(\varepsilon)>68.0005$, $0\le\varepsilon\le 0.011$. It follows that the vectors $\underline{\phi}$ of this paragraph are not optimal, the small upper bound on $\varepsilon$ being important to this conclusion.

The only remaining possibility for a fit that is better than $\underline{\psi}$ is a vector $\underline{\phi}\in\mathcal{R}^8$ that satisfies $\delta^2\phi_4\le 0$, $\delta^2\phi_5\le 0$, $\delta^2\phi_6\ge 0$ and $\delta^2\phi_7\ge 0$. We let $\underline{\phi}$ minimize $\sum_{i=1}^{8}(y_i-\phi_i)^2$ subject to these conditions. If $\delta^2\phi_5=0$ or $\delta^2\phi_6=0$ is achieved, then we know from the previous paragraph or the one before, respectively, that $\sum_{i=1}^{8}(y_i-\phi_i)^2>68.005$ occurs. Therefore we may assume that $\underline{\phi}$ has the properties $\delta^2\phi_5<0$ and $\delta^2\phi_6>0$. It follows that a small decrease in $\phi_5$, a small increase in $\phi_6$ and a small decrease in $\phi_7$ all preserve the constraints on $\underline{\phi}$. Thus we deduce $\phi_5\le 6-\varepsilon$, $\phi_6\ge 6-\varepsilon$ and $\phi_7\le 5$, but these conditions contradict $\delta^2\phi_6>0$. The proof that $\underline{\psi}\in\mathcal{R}^8$ is the unique best fit to the data of Table 2 when $k=3$ and $s=+1$ is complete.

Now the purpose of the example is to show that, although $\underline{\psi}\in\mathcal{R}^8$ is optimal, its first convex section and its concave section need not provide the best convex/concave fit to the data of these sections. Indeed, we see that the fit in Table 5 has the property that the only positive number in the sequence $\delta^2\phi_i$, $i=2,3,4,5$, comes before the only negative one. Further, the sum of squares of residuals has the value

$$\textstyle\sum_{i=1}^{6}(y_i-\phi_i)^2 = 68-12\,\varepsilon+41\,\varepsilon^2/\,77 = \sigma_3(\varepsilon), \qquad (4.10)$$

say, which satisfies the condition $\sigma_3(\varepsilon)<68$, $0<\varepsilon\le 0.011$. Therefore the example of this section does establish the fallibility of the conjecture that is assumed by the algorithms that are studied in Section 3.

Let Algorithm 3.1 be applied to the data of Table 2. Then equation (3.10) with $j=2$ yields the numbers $\gamma[3;2]=0$, $\gamma[6;2]=\sigma_3(\varepsilon)$, $\phi_{3;2}=4$ and $\phi_{6;2}=3-119\varepsilon/154$, because interpolation and Table 5 give the best convex/concave fits to the first 3 data and the first 6 data, respectively. We now consider the possibilities $q=3$ and $q=6$ in formula (3.8), the values of $k$, $(\alpha,\beta;k)$ and $n$ being 3, $\alpha$ and 8, respectively. When $q=3$, the quantity $\alpha[p,n]=\alpha[4,8]$ is the sum of squares (4.9), namely $\sigma_2(\varepsilon)$, that is derived from Table 4. Further $q=3$ is in the set $\mathcal{F}[n;k]$, because the relevant second difference

$$\begin{aligned}\delta^2\phi_4 &= \phi_{3;2}-10\,(5482-999\varepsilon)/1874+9\,(5700-956\varepsilon)/1874\\ &= (3976+1386\varepsilon)/1874 \qquad (4.11)\end{aligned}$$

| $i$ | $x_i$ | $y_i$ | $\phi_i$ | $\delta^2\phi_i$ | $y_i-\phi_i$ |
|---|---|---|---|---|---|
| 1 | 0 | $-6$ | $-6$ | | 0 |
| 2 | 1 | $-6$ | $-6$ | $9+13\varepsilon/154$ | 0 |
| 3 | 2 | 4 | $3+13\varepsilon/154$ | $-81-225\varepsilon/154$ | $1-13\varepsilon/154$ |
| 4 | 11 | $-4$ | $3-95\varepsilon/154$ | 0 | $-7+95\varepsilon/154$ |
| 5 | 12 | $6-\varepsilon$ | $3-107\varepsilon/154$ | 0 | $3-47\varepsilon/154$ |
| 6 | 13 | $6-\varepsilon$ | $3-119\varepsilon/154$ | | $3-35\varepsilon/154$ |

**Table 5** The best convex/concave fit to the first 6 data.

is positive. Therefore the integers $q(1)=1$ and $q(2)=3$ may occur in equation (3.2), so $\gamma[3;2]+\sigma_2(\varepsilon)=\sigma_2(\varepsilon)$ is an upper bound on the final value of $\sum_{i=1}^{n}(y_i-\phi_i)^2$. On the other hand, the integer $q=6$ is not in the set $\mathcal{F}[n;k]$, because $\alpha[p,n]=\alpha[7,8]=0$ is achieved by interpolation, so $\delta^2\phi_7$ has the value

$$\delta^2\phi_7 = 20\,\phi_{6;2} - 21\times 5 + 5 = -40 - 1190\,\varepsilon/77, \tag{4.12}$$

which is negative. Thus $q(2)=6$ is excluded from the sum (3.2). It follows that Algorithm 3.1 fails to provide the optimal fit $\underline{\psi}$. Instead, by choosing $q(1)=1$ and $q(2)=3$, the algorithm yields a feasible vector $\underline{\phi}$ that has the sum of squares of residuals

$$\textstyle\sum_{i=1}^{n}(y_i-\phi_i)^2 = \sigma_2(\varepsilon) \approx 68.13 - 11.60\,\varepsilon + 1.00\,\varepsilon^2, \tag{4.13}$$

which is very close to the optimal value (4.2).

# 5 A new algorithm

We will continue to take advantage of the main result of Section 2, namely that the optimal fit $\underline{\phi}\in\mathcal{R}^n$ can be divided into convex and concave sections, where each section can be calculated straightforwardly from the indices of its end-points. Furthermore, we take the view that the purpose of formula (3.10) is to answer the following question. If $\bar{q}$ is the index of the right hand end-point of the $j$-th section, what is the index of the right hand end-point of the $(j-1)$-th section? Specifically, the algorithms of Section 3 assume that the required index is the integer $q$ that provides the least sum on the right hand side of expression (3.10). It is shown in Section 4, however, that this assumption may be incorrect. Indeed, we find in Table 2 that, when $j=2$ and $\bar{q}=6$, the answer should be the integer $q=4$, but the fit in Table 5 shows that formula (3.10) actually gives priority to the choice $q=2$. Therefore we seek a more suitable way of choosing $q$, that will provide a reliable dynamic programming algorithm for our data fitting problem.

Now the essence of the dynamic programming method is that, given $\bar{q}$ and $j$, the selection of $q$ in the previous paragraph should not make use of any of the data $(x_i, y_i)$, $i=\bar{q}+1, \bar{q}+2, \ldots, n$. In the case of the example with $j=2$ and $\bar{q}=6$, however, the choice between $q=2$ and $q=4$ does depend on these subsequent data. Therefore, instead of generating a single value of $q$ by formula (3.10), we require a technique that provides all possible values of $q=q(j-1)$ in the trial sum (3.2), when we are considering the assumption $q(j)=\bar{q}$. Furthermore, the decisions of the technique have to depend only on $(x_i, y_i)$, $i=1,2,\ldots,\bar{q}$.

Given these data, where $j$ and $\bar{q}$ are any integers from the intervals $[2,k]$ and $[j,n]$, respectively, we seek a small subset of the integers $\{j-1, j, \ldots, \bar{q}-1\}$ such that, if $q(j)=\bar{q}$ occurs in the final fit, then $q(j-1)$ is an element of the subset. Further, for each $q$ in the subset, we store numbers that correspond to $\gamma[\bar{q};j]$ and $\phi_{\bar{q};j}$ in expression (3.11). We show the dependence on $q$ by calling the new numbers $\gamma[q,\bar{q};j]$ and $\phi_{q,\bar{q};j}$. Specifically, $\gamma[q,\bar{q};j]$ is the value of $\sum_{i=1}^{\bar{q}}(y_i-\phi_i)^2$ that would be achieved by the optimal fit in the case $q(j-1)=q$ and $q(j)=\bar{q}$, while $\phi_{q,\bar{q};j}$ is the element $\phi_{\bar{q}}$ of that fit. Therefore $\phi_{q,\bar{q};j}$ is just the component $\phi_{\bar{q}}$ of the solution of the quadratic programming problem (2.3), when the integers $p$, $q$ and $j$ of expression (2.3) are the current values of $q+1$, $\bar{q}$ and $j$, respectively. That problem also defines all of the components $\phi_i$, $i=q+1, q+2, \ldots, \bar{q}$, which provide a barrier between $\phi_i$, $i \le q$, and $\phi_i$, $i \ge \bar{q}+1$. It follows that the value of $\gamma[q,\bar{q};j]$ is independent of the data $(x_i, y_i)$, $i=\bar{q}+1, \bar{q}+2, \ldots, n$, so it can be calculated by the dynamic programming method.

Now an important difference between Algorithm 3.1 and the methods of Demetriou (1985) is that, if $q(j)=\bar{q}$ and $\bar{q}<n$, then we employ the condition

$$(-1)^{j+1}\,\phi[x_{\bar{q}}, x_{\bar{q}+1}, x_{\bar{q}+2}] \le 0 \quad \text{or} \quad q(j+1) = q(j)+1 \qquad (5.1)$$

to ensure that the $(j+1)$-th section of $\underline{\phi}$ is joined to the $j$-th section of $\underline{\phi}$ in a way that provides feasibility. This condition is highly suitable for the new algorithm of this section, because the information that is required from the calculation of the previous paragraph is only the value of $\phi_{q,\bar{q};j}$. Moreover, if inequality (5.1) holds when $q=\ell$ in the equation $\phi_{\bar{q}} = \phi_{q,\bar{q};j}$, then it is satisfied by all the integers $q$ in $[j-1, \bar{q}-1]$ that have the property $(-1)^{j+1}\phi_{q,\bar{q};j} \le (-1)^{j+1}\phi_{\ell,\bar{q};j}$. Therefore there is no need to retain $\ell$ if one of these integers $q$ also achieves the condition $\gamma[q,\bar{q};j] \le \gamma[\ell,\bar{q};j]$, because then the choice $q(j-1)=q$ is never worse than the choice $q(j-1)=\ell$.

Therefore the subset of the integers $\{j-1, j, \ldots, \bar{q}-1\}$ that is mentioned in the paragraph before last, $\mathcal{S}[\bar{q};j]$ say, is formed as follows. We let $\mathcal{S}[\bar{q};j]$ be empty initially, and then we let $q$ run through the integers $j-1, j, \ldots, \bar{q}-1$. For each $q$, we calculate the numbers $\gamma[q,\bar{q};j]$ and $\phi_{q,\bar{q};j}$ that have been mentioned already, except that some values of $q$ may have been excluded for a reason that will be given later. If $q$ is not excluded, then we add it to $\mathcal{S}[\bar{q};j]$, unless the current $\mathcal{S}[\bar{q};j]$ contains an integer $\ell$ that satisfies the inequalities

$$\gamma[\ell,\bar{q};j] \le \gamma[q,\bar{q};j] \quad \text{and} \quad (-1)^{j+1}\phi_{\ell,\bar{q};j} \le (-1)^{j+1}\phi_{q,\bar{q};j}. \tag{5.2}$$

Further, if $q$ is added to $\mathcal{S}[\bar{q};j]$, then we delete from $\mathcal{S}[\bar{q};j]$ every integer $\ell \neq q$ that has the properties

$$\gamma[q,\bar{q};j] \le \gamma[\ell,\bar{q};j] \quad \text{and} \quad (-1)^{j+1}\phi_{q,\bar{q};j} \le (-1)^{j+1}\phi_{\ell,\bar{q};j}, \tag{5.3}$$

at least one of these inequalities being strict, because otherwise $q$ would not have become an element of $\mathcal{S}[\bar{q};j]$. This set corresponds to the integer $q[\bar{q};j]$ in Algorithm 3.1. Therefore, for each integer $\bar{q}$ in $[3,n-1]$, the new algorithm stores the numbers

$$\gamma[\bar{q};1],\ \phi_{\bar{q};1} \quad \text{and} \quad \{\gamma[q,\bar{q};j],\, \phi_{q,\bar{q};j} : q\in\mathcal{S}[\bar{q};j]\}, \quad j=2,3,\ldots k(\bar{q}), \tag{5.4}$$

instead of the values (3.11). We expect the cardinality of each $\mathcal{S}[\bar{q};j]$ to be small, but this question has not been investigated carefully.

We require some more notation in order to describe the analogue of equation (3.10) for calculating $\gamma[q,\bar{q};j]$. It is convenient to write $\gamma[0,\bar{q};1]$ and $\phi_{0,\bar{q};1}$ instead of $\gamma[\bar{q};1]$ and $\phi_{\bar{q};1}$, and also we let each of the sets $\mathcal{S}[\bar{q};1]$, $\bar{q}=1,2,\ldots,n$, contain just the integer 0. Further, we let $\mathcal{F}[q,\bar{q};j]$ be the subset of $\mathcal{S}[q;j-1]$ such that, if $\ell\in\mathcal{F}[q,\bar{q};j]$, then feasibility allows the $(j-1)$-th section of the fit to run from $x_{\ell+1}$ to $x_q$ and the $j$-th section of the fit to run from $x_{q+1}$ to $x_{\bar{q}}$, where $\bar{q}$, $j$ and $q$ are any integers from the intervals $[3,n]$, $[2,k(\bar{q})]$ and $[j-1,\bar{q}-1]$, respectively. Therefore we include the integer $\ell$ in $\mathcal{F}[q,\bar{q};j]$ if and only if it is an element of $\mathcal{S}[q;j-1]$ that is allowed by the feasibility condition

$$(-1)^j\,\phi[x_q,x_p,x_{p+1}] \le 0 \quad \text{or} \quad \bar{q}=q+1. \tag{5.5}$$

Here $p=q+1$ as usual, $\phi_q$ has the value $\phi_{\ell,q;j-1}$, and $\phi_p$ and $\phi_{p+1}$ occur in the fit that gives the sum of squares $(\alpha,\beta;j)[p,\bar{q}]$. If $\mathcal{F}[q,\bar{q};j]$ is empty, then $q$ is excluded from the new set $\mathcal{S}[\bar{q};j]$, as mentioned near the beginning of the previous paragraph. This situation is demonstrated by the example $y_i=(x_i-x_1)^2$, $i=1,2,3$, and $y_i=-(x_6-x_i)^2$, $i=4,5,6$, that is considered after equation (1.4), but now we assume $n>6$ and $k\ge 3$. Specifically, we omit $q=3$ from $\mathcal{S}[6;2]$, because the set $\mathcal{F}[3,6;2]$ is empty. Indeed, $\mathcal{S}[3;1]=\{0\}$ implies that $\ell=0$ is the only possible element of $\mathcal{F}[3,6;2]$. When $\ell=0$, $q=3$ and $\bar{q}=6$, however, then the values $\phi_3=\phi_{0,3;1}=y_3$, $\phi_4=y_4$ and $\phi_5=y_5$ occur in condition (5.5), which prohibits $\ell=0$, because $j$ is even and $\phi[x_3,x_4,x_5]=y[x_3,x_4,x_5]$ is positive.

Therefore, for all combinations of the integers $\bar{q}\in[3,n]$ and $j\in[2,k(\bar{q})]$, where $\bar{q}$ is adjusted in the outermost loop, the set $\mathcal{S}[\bar{q};j]$ is formed. Then, for each $q\in\mathcal{S}[\bar{q};j]$, the formula

$$\gamma[q,\bar{q};j] = \min\{\gamma[\ell,q;j-1] + (\alpha,\beta;j)[p,\bar{q}] : \ell\in\mathcal{F}[q,\bar{q};j]\} \tag{5.6}$$

is applied, where $p=q+1$, and where the numbers $\gamma[\ell,q;j-1]$ are available already by recursion, because the algorithm employs the starting values

$$\left.\begin{array}{l}\gamma[0,1;1]=\gamma[0,2;1]=\gamma[1,2;2]=0, \quad \phi_{0,1;1}=y_1,\\ \phi_{0,2;1}=\phi_{1,2;2}=y_2 \quad \text{and} \quad \mathcal{S}[2;2]=\{1\},\end{array}\right\} \tag{5.7}$$

instead of the values (3.12). Thus Steps 0–4 of the new algorithm are analogous to Steps 0–4 of Algorithm 3.1, the only changes being that we replace expressions (3.12), (3.11) and (3.10) by expressions (5.7), (5.4) and (5.6), respectively.

The new version of formula (3.9) sets $q(k)=n$ as before, and then it has to generate indices $q(j-1)$, $j=k,k-1,\ldots,2$, that make $\underline{\phi}\in\mathcal{R}^n$ feasible, and that cause expression (3.2) to be the sum of squares

$$\textstyle\sum_{i=1}^{n}(y_i-\phi_i)^2=\min\{\gamma[q,n;k]:q\in\mathcal{S}[n;k]\}. \tag{5.8}$$

Therefore $q(k-1)$ is set to the value of $q$ that provides the minimum on the right hand side of this equation. Further, it follows from formula (5.6) that $q(k-2)$ is the value of $\ell$ in the set $\mathcal{F}[q(k-1),n;k]$ that is defined by the identity

$$\gamma[q(k-1),n;k]=\gamma[\ell,q(k-1);k-1]+(\alpha,\beta;k)[q(k-1)+1,n]. \tag{5.9}$$

These remarks provide the choices of $q(k)$, $q(k-1)$ and $q(k-2)$ that are made by the new algorithm. Further, if $k\geq 4$, then for $j=k-1,k-2,\ldots,3$, the value of $q(j-2)$ is set to the integer $\ell$ in $\mathcal{F}[q(j-1),q(j);j]$ that has the property

$$\gamma[q(j-1),q(j);j]=\gamma[\ell,q(j-1);j-1]+(\alpha,\beta;j)[q(j-1)+1,q(j)]. \tag{5.10}$$

Next the integers $p(j)$, $j=1,2,\ldots,k$, are provided by expression (3.1) as before. Finally, the new algorithm applies Step 6 of Algorithm 3.1.

The success of the new method is guaranteed, because there are no assumptions or conditions that exclude the required best fit. Further, we pick the least value of $\sum_{i=1}^{n}(y_i-\phi_i)^2$ that is allowed by the constraints of the calculation, and the constraints ensure that the final $\underline{\phi}$ is feasible. We note, however, as indicated at the beginning of Section 3, that the final fit need not have a convex section on $[x_{p(j)},x_{q(j)}]$ when $j$ is odd or a concave section on this interval when $j$ is even. For example, because the elements of the set $\mathcal{S}[\bar{q};j]$ are all less than $\bar{q}$, the new algorithm would not fit the data $y_i=(x_i-x_1)^2$, $i=1,2,\ldots,n$, by a single section when $k\geq 2$. Instead, it can calculate the values $p(1)=1$, $q(1)=n-k+1$ and $p(j)=q(j)=n-k+j$, $j=2,3,\ldots,k$, which also provide $\phi_i=y_i$, $i=1,2,\ldots,n$. This anomaly is permitted by condition (5.5), because we see that the sign of the second difference is irrelevant in the case $q=q(j)$ and $\bar{q}=q(j)+1=q(j+1)$.

Another important feature of condition (5.5) is that it does not depend on $\phi_{q-1}$. This property is highly advantageous to the new algorithm, because the task of joining two sections together in a feasible way would be far more difficult if one had to consider four consecutive components of $\underline{\phi} \in \mathcal{R}^n$ instead of only three. Here we are taking advantage of the remark that, if the differences $\phi[x_{q-2}, x_{q-1}, x_q]$ and $\phi[x_q, x_p, x_{p+1}]$ have opposite signs, where $p = q+1$, then the sign of $\phi[x_{q-1}, x_q, x_p]$ is irrelevant to feasibility. Occasionally this device allows some unexpected freedom in the calculated integers $p(j)$ and $q(j)$, $j = 1, 2, \ldots, k$. For example, if $k = 2$, if the first $\ell - 1$ of the differences $y[x_{i-1}, x_i, x_{i+1}]$, $i = 2, 3, \ldots, n-1$, are strictly positive, and if the remaining $n - \ell - 1$ differences are strictly negative, then the theory of Section 2 suggests that $q(1) = \ell$ and $p(2) = \ell + 1$ will occur. Our method can calculate the best fit, however, by picking the values $q(1) = \ell + 1$ and $p(2) = \ell + 2$, because the sign of $\phi[x_{q(1)-1}, x_{q(1)}, x_{p(2)}]$ is ignored. Such behaviour occurs in the calculated fit to the data only if some suitable interpolation conditions are satisfied at a join.

All of the work that is presented in Sections 4 and 5 was done after the conference in July. Consequently, there has not been enough time to run any numerical experiments, in order to test the efficiency of the new algorithm.

# References

[1] Cullinan, M.P. (1986) *Data Smoothing by Adjustment of Divided Differences*, Ph.D. Dissertation, University of Cambridge.

[2] Demetriou, I.C. (1985) *Data Smoothing by Piecewise Monotonic Divided Differences*, Ph.D. Dissertation, University of Cambridge.

[3] Demetriou, I.C. and M.J.D. Powell (1991) "The minimum sum of squares change to univariate data that gives convexity", *IMA J. Numer. Anal.* **11**, 433–448.

[4] Hildebrand, F.B. (1956) *Introduction to Numerical Analysis,* McGraw-Hill Book Company (New York).

# A Framework for Interpolation and Approximation on Riemannian Manifolds

*Nira Dyn*

School of Mathematical Sciences, Tel Aviv University, Tel Aviv 69978, Israel

*Francis J. Narcowich*
*Joseph D. Ward*

Department of Mathematics, Texas A&M University, College Station, TX 77843, USA[1]

**Abstract**

In this paper we provide a framework for studying the approximation order resulting from using strictly positive definite kernels to do generalized Hermite interpolation and approximation on a compact Riemannian manifold. We apply this framework to obtain explicit estimates in the cases of the circle and 2-sphere. In addition, we provide a technique for constructing strictly positive definite spherical functions out of radial basis functions, and we use it to make a spherical function that is locally supported.

# 1 Introduction

## 1.1 Overview

The object of this paper is to provide a brief overview of our recent investigations concerning variational principles and Sobolev-type estimates for the approximation order resulting from using strictly positive definite kernels to

[1]Research of the second and third authors sponsored by the Air Force Office of Scientific Research, Air Force Materiel Command, USAF, under grant number F49620-95-1-0194. The U.S. Government is authorized to reproduce and distribute reprints for Governmental purposes notwithstanding any copyright notation thereon. The views and conclusions contained herein are those of the authors and should not be interpreted as necessarily representing the official policies or endorsements, either expressed or implied, of the Air Force Office of Scientific Research or the U.S. Government.

Research of the third author supported in part by National Science Foundation Grant DMS-9505460.

do generalized Hermite interpolation on a closed (i.e., no boundary), compact, connected, orientable, $m$-dimensional $C^\infty$ Riemannian manifold $M^m$. A complete discussion of the results stemming from these investigations may be found in [1]. The manifolds of particular interest in this article will be the circle and the 2-sphere, although our techniques apply to certain manifolds which arise in group theory as well. We point out that in the case of the 2-sphere some work in this direction has already been done [3, 6, 11, 14, 15]. The rates of approximation discussed here will be analyzed in terms of Sobolev norms. We point out that the constants appearing in the approximation order inequalities are explicit. (See §5.1 and §5.2). Finally, we mention that in §3.3 we give some new results that allow one to make make use of radial basis functions (RBFs) on $R^{m+1}$ to obtain strictly positive definite functions on $S^{m-1}$. We use this to take a compactly supported RBF in $R^5$ [16] and produce a locally supported, strictly postive definite function on $S^m$, $m \leq 3$.

## 1.2 Sobolev spaces

In this section we briefly describe the notion of a Sobolev space on a $C^\infty$ manifold; for proofs and more references, see [1, 8].

The Riemannian metric $g^{ij}$ on $M^m$ induces the standard measure, $d\mu$, on $M^m$, and also gives rise to the Laplace–Beltrami operator, $\Delta$, which is elliptic and self-adjoint relative to $L^2(M^m, g)$. The spectrum of $-\Delta$ is discrete, starts at 0, and has $+\infty$ as its only accumulation point; each eigenvalue in this spectrum has finite multiplicity. The eigenfunctions corresponding to these eigenvalues are $C^\infty$ functions. We will label the eigenvalues the with index $j$ that comes from listing the eigenvalues in increasing order $\lambda_0 = 0 < \lambda_1 \leq \lambda_2 \ldots$, with appropriate repetitions for degenerate cases. We will let $F_j$ be the eigenfunction corresponding to the eigenvalue $\lambda_j$, and we will choose the $F_j$'s so that they form an orthonormal basis for $L^2(M^m, g)$.

The notion of a distribution may be defined on a manifold; as usual it is a linear functional defined on $C^\infty(M^m)$. We shall denote the set of all distributions on the manifold $M^m$ by $\mathcal{D}'(M^m)$. The action of distributions on test functions will be given by

$$(u, \phi) = \int_{M^m} u(p)\phi(p)d\mu(p). \tag{1.1}$$

Since the eigenfunctions $F_j$ are themselves $C^\infty$ functions, we may compute the coefficients in a formal eigenfunction expansion for $u$:

$$\widehat{u}(j) := (u, \bar{F}_j)\,.$$

The eigenfunction expansion for $u$ can be used to express (1.1) in the form [8, pp. 178–9]

$$(u, \bar{\phi}) = \sum_{j=0}^{\infty} \widehat{u}(j)\overline{\widehat{\phi}(j)}. \tag{1.2}$$

The Sobolev space associated with an arbitrary real number $s$ is

$$H_s(M^m) = \{u \in \mathcal{D}'(M^m) : \sum_{j=1}^{\infty} \lambda_j^s |\hat{u}(j)|^2 < \infty\},$$

$$\text{with norm } \|u\|_s = \left( \sum_{j=0}^{\infty} (1+\lambda_j)^s |\hat{u}(j)|^2 \right)^{1/2}.$$

The Sobolev embedding theorem applies here, and so $C^k(M^m) \supset H_s(M^m)$ whenever $s > k + \frac{m}{2}$. In addition, the standard duality $H_{-s} = H_s^*$ holds.

# 2 Positive definite kernels

A continuous, complex-valued kernel $\kappa(\cdot,\cdot)$ is termed *positive definite on $M^m$* if $\bar{\kappa}(q,p) = \kappa(p,q)$ and if for every finite set of points $C = \{p_1, \ldots, p_n\}$ in $M^m$, the self-adjoint, $N \times N$ matrix with entries $\kappa(p_j, p_k)$ is positive semi-definite. (See [13, 8, 1].)

One can view this positivity in terms of distributions. If $\delta_p$ is the Dirac delta function located at $p$ and if $u = \sum_j c_j \delta_{p_j}$, then the matrix $[\kappa(p_j, p_k)]$ being positive semi-definite is equivalent to the quadratic form $(\bar{u} \otimes u, \kappa) \geq 0$ for every choice of $c_j$'s. This result extends to a somewhat more general situation. By [1, Proposition 1.3], if $u$ and $v$ are distributions in $H_{-s}(M^m)$, then the tensor product $u \otimes v$ is in $H_{-2s}(M^m \times M^m)$. It follows that when $\kappa$ is in $H_{2s}(M^m \times M^m)$, $(u \otimes v, \kappa)$ is well defined, and by [1, Theorem 2.1], $(\bar{u} \otimes u, \kappa) \geq 0$. Thus $\kappa$ is positive definite on $H_{-s}(M^m)$. This "extension theorem" has a corollary that is useful for constructing positive definite kernels on manifolds embedded in $R^n$ (or any other manifold, for that matter).

**Corollary 2.1** *If $\kappa(q,p)$ is a continuous positive definite kernel on $R^n$, and if $M^m$ is embedded in $R^n$, then the restriction of $\kappa$ to $M^m$ is positive definite.*

**Proof** The matrix $[\kappa(p_j, p_k)]$ is positive semi-definite for any distinct set of points in $R^n$. It will therefore be positive definite when the points are restricted to $M^m$. □

The mapping properties associated with a kernel $\kappa \in H_{2s} \cap C^0(M^m \times M^m)$ are important to us. Define the linear transformation $\Psi_\kappa : C^0(M^m) \to C^0(M^m)$ via the expression

$$\Psi_\kappa[f](p) = \int_{M^m} \kappa(p,q) f(q) d\mu(q).$$

In [1, Proposition 1.3], we show that $\Psi_\kappa$ is a bounded linear map from $H_{-s}(M^m)$ into $H_s(M^m)$. To emphasize the role the kernel plays here, we will write

$$\kappa \star f(p) := \psi_\kappa[f](p).$$

# 3 Strictly positive definite kernels

## 3.1 Generalized Hermite interpolation

The interpolation problem that we wish to address is the following. Think of a function $f$ that is smooth, say in $H_s$, and that we wish to reconstruct from data whose measurement is mathematically modeled as the application of a linear functional to the unknown function $f$. For example, measuring the value of $f$ at a point $p$ can be thought of as obtaining $(\delta_p, f)$. More generally, take $\mathcal{U} := \{u_j\}_{j=1}^N \subset H_{-s}$ to be a linearly independent (i.e., non-redundant) set of distributions on $M^m$, and take the measurements to be

$$\int_{M^m} \bar{u}_j(p) f(p) d\mu(p) = d_j, \quad \text{for } j = 1, \ldots, N, \tag{3.1}$$

where the $d_j$'s are complex numbers. (We remark that the complex conjugate appears above to simplify formulas used later on.)

The choice of interpolant requires discussion. In the case of scattered-data interpolation on $R^m$ with an order-0 radial basis function $F$ (see [10] for a review), the interpolant has the form

$$\begin{aligned}
\tilde{f}(x) &= \sum_{j=1}^{N} c_j F(\|x - \xi_j\|) \\
&= \sum_{j=1}^{N} c_j \left(\delta_{\xi_j}(y), F(\|x - y\|)\right) \\
&= \Big(\underbrace{\sum_{j=1}^{N} c_j \delta_{\xi_j}(y)}_{u(y)}, F(\|x - y\|)\Big) \\
&= (u(y), F(\|x - y\|)) \\
&= F(\|x - \cdot\|) \star u\,.
\end{aligned}$$

This suggests that we look for an interpolant of the form

$$\tilde{f} := \kappa \star u \text{ where } u \in \mathcal{U}. \tag{3.2}$$

## 3.2 Strict positivity and an inner product for $H_{-s}$

Now that we have settled on the form of the interpolant, we need to find the kernels for which one always has that there is a $u \in \mathcal{U}$ such that (3.1) is satisfied by $\tilde{f}$ in (3.2). By standard arguments, this is equivalent to the invertibility of the interpolation matrix $A$ with entries

$$\begin{aligned}
A_{j,k} &= (\bar{u}_j, \kappa \star u_k) \\
&= (\bar{u}_j \otimes u_k, \kappa).
\end{aligned}$$

The quadratic form associated with $A$ is

$$c^*Ac = (\bar{u} \otimes u, \kappa) \text{ where } u = \sum_{j=1}^{N} c_j u_j. \tag{3.3}$$

If $\kappa$ is positive definite, then, by our earlier remarks in §2, $c^*Ac \geq 0$; thus, $A$ is positive semi-definite. This is still not sufficient for $A$ to be invertible under all circumstances. For that, we will require that the kernel $\kappa$ satisfy

$$(\bar{u} \otimes u, \kappa) > 0 \quad \forall\, u \neq 0,\ u \in H_{-s}(M^m). \tag{3.4}$$

If (3.4) holds, we will say that $\kappa$ is *strictly positive definite.* Since a strictly positive definite kernel means a positive definite interpolation matrix for any of the interpolation problems described in §3, all of these problems are well posed.

One can use strictly positive definite kernels that are continuous and in $H_{2s}(M^m \otimes M^m)$ to define an inner product on $H_{-s}$; simply let

$$[\![u, v]\!] := (\bar{v} \otimes u, \kappa) \tag{3.5}$$
$$|\!|\!|u|\!|\!| := (\bar{u} \otimes u, \kappa)^{1/2}. \tag{3.6}$$

We note that although one could complete $H_{-s}$ to a Hilbert space using the inner product (3.5), it is not necessary to do so.

Strictly positive definite kernels have the property that all of the matrices $[\kappa(p_j, p_k)]$ are positive definite for an arbitrary finite set of distinct points $\{p_1, \ldots, p_N\}$. It is tempting to think that the converse is true. Sad to say, there is a counterexample; see the paper by Ron and Sun [11].

## 3.3 Examples

Our basic source of examples is the following theorem.

**Theorem 3.1** *Let $\kappa(p, q)$ be a continuous kernel in $H_{2s}(M^m \times M^m)$ having the eigenfunction expansion*

$$\kappa(p, q) := \sum_{\alpha \in \mathcal{A}} a(\alpha) F_\alpha(p) \bar{F}_\alpha(q), \tag{3.7}$$

*where $\mathcal{A}$ is some countable index set, and the $F_\alpha$'s are the normalized eigenfunctions of the Laplace–Beltrami operator on $M^m$. $\kappa$ is strictly positive definite if and only if $a(\alpha) > 0$ for all $\alpha \in \mathcal{A}$.*

**Corollary 3.2** *If the strictly positive definite kernels $\kappa_1 \ldots, \kappa_n$ all have the form (3.7) on $M^{m_1}, \ldots, M^{m_n}$, then $\kappa = \kappa_1 \cdots \kappa_n$ is strictly positive definite on the product manifold $M^{m_1} \times \cdots \times M^{m_n}$.*

Here are a few explicit examples of such kernels for various manifolds.

1. ($M^m$) $\kappa_t(p,q) = \sum_{j=1}^{\infty} \exp(-t\lambda_j) F_j(p)\bar{F}_j(q)$, the heat kernel on $M^m$.

2. ($S^m$) $\kappa(p,q) = \sum_{\ell,j} a(\ell,j) Y_{\ell,j}\bar{Y}_{\ell,j}$, where the $Y_{\ell,j}$ are the spherical harmonics on $S^m$.

3. ($S^m$) $\kappa(p,q) = \sum_{j=0}^{\infty} b_j \cos^j(\theta(p,q))$, $b_j > 0$, where $\theta(p,q)$ denotes the length of the shorter of the two arcs on the great circle joining $p$ and $q$. The dimension $m$ can be arbitrary.

4. ($S^m$) $\kappa(p,q) = \exp(s\cos(\theta(p,q)))$, $s > 0$ ($m$ arbitrary).

5. ($S^3$) $\kappa(p,q) = \left(1 - 2\sin(\frac{\theta(p,q)}{2})\right)_+^5 \left(10\sin(\frac{\theta(p,q)}{2}) + 1\right)$ (see Remark 3.4).

6. ($T^m$) $\kappa(\phi,\phi') = K(\phi - \phi') = \sum_{\alpha \in Z^m} a(\alpha) e^{i\alpha\cdot(\phi-\phi')}$.

7. ($T^m$) $K(\phi) = \exp(\sum_{j=1}^{m} s(j)\cos(\phi_j))$, $s_j > 0$.

8. ($T^m$) $K(\phi) = \prod_{1\le j\le m} g(\phi_j)$, where $g$ is an even, $2\pi$-periodic function. On $[0,\pi]$, $g$ is defined by

$$g(\phi) = \begin{cases} a^{-5}(a-\phi)^3(a^2 + 3a\phi + \phi^2) & \text{if } 0 \le x \le a \\ 0 & \text{if } a < \phi \le \pi. \end{cases}$$

Let $f(x) = F(\|x\|)$ be a smooth, order-0 RBF defined on $R^{m+1}$. If $q$ and $p$ are points on the unit $S^m$ embedded in $R^{m+1}$, we recall that

$$\|p - q\|^2 = 2 - 2p\cdot q = 2 - 2\cos(\theta),$$

where $p \cdot q$ is the Euclidean dot-product and $\theta$ is the angle between $p$ and $q$. Define the function

$$G(\cos(\theta)) := F\left(\sqrt{2 - 2\cos(\theta)}\right). \tag{3.8}$$

By virtue of Corollary 2.1, $G(p\cdot q) = F(\|p-q\|)$ is a positive definite kernel on $S^m$. Moreover, since $G$ is a function of $\cos(\theta)$ alone, it is also a positive definite spherical *function* [12], and so admits an expansion in Legendre polynomials [7, p. 21] appropriate to $S^m$,

$$G(\cos(\theta)) = \sum_{\ell=0}^{\infty} g_m(\ell) P_\ell(m+1; \cos(\theta)) \tag{3.9}$$

with $g_m(\ell) \ge 0$. By [9, Proposition 2.1], if $g_m(\ell) > 0$ for infinitely many even $\ell$ and infinitely many odd $\ell$, then it is strictly positive definite on $S^{m-1}$ Alternatively, if we let $t = \cos(\theta)$, then neither the even part nor the odd part of $G(t)$ is a polynomial. We summarize these remarks below.

**Corollary 3.3** *If $F$ is an order-0 RBF on $R^{m+1}$, then $G$ is a positive definite function on $S^m$. Moreover, if neither of the two functions,*

$$G(\cos(\theta)) \pm G(-\cos(\theta))$$

*is a polynomial in* $\cos(\theta)$*, then $G$ is strictly positive definite on $S^{m-1}$.*

This corollary allows us to draw upon some of the new, compactly supported RBFs that Wendland [16] has constructed. Although these RBFs have the slight disadvantage of being dimension dependent, their having compact support more than makes up for it. One of these, the RBF below, is positive definite on $R^5$ and is $C^2$.

$$\phi_{5,1}(r) := (1-r)_+^5(5r+1)\,. \tag{3.10}$$

Using the prescription above, we first let $r = \sqrt{2-2\cos(\theta)}$, and then set

$$\begin{aligned}\gamma(\cos(\theta)) &:= \phi_{5,1}\left(\sqrt{2-2\cos(\theta)}\right) \\ &= \left(1-\sqrt{2-2\cos(\theta)}\right)_+^5 (1+5\sqrt{2-2\cos(\theta)}) \\ &= \left(1-2\sin\left(\frac{\theta}{2}\right)\right)_+^5 \left(10\sin\left(\frac{\theta}{2}\right)+1\right).\end{aligned} \tag{3.11}$$

**Remark 3.4** *The function* $\gamma(\cos(\theta))$ *defined above is strictly positive definite on $S^m$ for $m = 1$, 2, and 3.*

**Proof** Observe that $r = \sqrt{2-2\cos(\theta)} = 2\sin\left(\frac{\theta}{2}\right)$ is larger than 1 for $\pi/3 \le \theta \le \pi$. Hence, $\gamma(\cos(\theta)) = 0$ for such $\theta$. On the other hand, replacing $\cos(\theta)$ by $-\cos(\theta)$ gives us $r = \sqrt{2+2\cos(\theta)} = 2\cos\left(\frac{\theta}{2}\right)$, which exceeds 1 on the interval $0 \le \theta \le 2\pi/3$. Hence, $\gamma(-\cos(\theta)) = 0$ on the interval $0 \le \theta \le 2\pi/3$. It follows that the even and odd parts of $\gamma(\cos(\theta))$ vanish identically for $\pi/3 \le \theta \le 2\pi/3$, and so neither can be polynomials in $\cos(\theta)$. Of course, $\gamma$ is not identically 0; thus, $\gamma$ is strictly positive definite on $S^m$ with $m \le 5-2=3$. □

# 4 A variational framework for interpolation

## 4.1 The variational principle and basic bounds

We will assume at the outset that $\kappa$ is strictly positive definite, continuous, and in $H_{2s}(M^m \times M^m)$. The generalized Hermite interpolation problem described in §3.1, equation (3.1), can be recast in terms of the inner product defined by (3.5). We assume that the data is generated by applying distributions from the set $\{u_j\}_{j=1}^N \subset H_{-s}$ to an unknown $H_s(M^m)$ function of the form $f = \kappa \star v$, with $v$ in $H_{-s}(M^m)$. That is, we assume the data set has the form

$$(\bar{u}_j, f) = [\![v, u_j]\!] = d_j \quad \text{for } j = 1, \dots, N. \tag{4.1}$$

Let $\mathcal{U} = \text{Span}\left(\{u_j\}_{j=1}^N\right)$. We seek to interpolate the data and approximate the function $f = \kappa\star v$ with a function of the form $\tilde{f} = \kappa\star u$ for some $u \in \mathcal{U}$. As we mentioned earlier, the interpolation matrix for this problem is invertible, and so there exists a unique $u \in \mathcal{U}$ that solves it. The distribution $u$ satisfies the following minimization principle.

**Theorem 4.1** [1, Theorem 3.1] *Let $u \in H_{-s}(M^m)$ be the unique distribution in $\mathcal{U}$ for which $\kappa \star u$ satisfies*

$$[\![u, u_j]\!] = d_j \quad \textit{for } j = 1, \dots, N.$$

*If $v \in H_{-s}(M^m)$ solves (4.1) then $v - u$ is orthogonal to $\mathcal{U}$ with respect to the inner product (3.5). In addition,*

$$\|v\|^2 = \|v - u\|^2 + \|u\|^2.$$

*Finally, if $v \neq u$,*

$$\|u\| < \|v\|.$$

We remark that this is similar to the variational principle derived by Madych and Nelson [4, 5] in the radial basis function case. The last inequality in the theorem amounts to saying that among all $v \in H_{-s}(M^m)$ for which $\kappa\star v$ interpolates the data, the distribution $u$ minimizes the norm (3.6). Since both $f$ and $\tilde{f}$ satisfy (4.1), one has that

$$(\bar{u}_j, f - \tilde{f}) = [\![v - u, u_j]\!] = 0 \text{ for } j = 1, \dots, N,$$

and so $v - u$ is in $\mathcal{U}^\perp$. Clearly, $u$ is the orthogonal projection of $v$ onto $\mathcal{U}$. Next, let $w$ be an arbitrary distribution in $H_{-s}(M^m)$. We wish to bound the quantity $(\bar{w}, f - \tilde{f})$ ; such a bound is useful in getting rates of approximation. Rewriting this quantity as

$$(\bar{w}, f - \tilde{f}) = [\![v - u, w]\!],$$

and using the fact that $v - u \in \mathcal{U}^\perp$ together with standard Hilbert space methods, we obtain [1, Proposition 3.2]

$$|(\bar{w}, f - \tilde{f})| \leq \text{dist}(v, \mathcal{U})\, \text{dist}(w, \mathcal{U})\,. \tag{4.2}$$

The utility in (4.2) is that we can estimate the effect of $w$ applied to $f - \tilde{f}$ independently of the original $f$. Also, the distributions $w$ and $v$ appear symmetrically on the right-hand side above, in factors that involve only distances to $\mathcal{U}$. Thus our strategy for obtaining estimates on $|(\bar{w}, f - \tilde{f})|$ is to get upper bounds on the distance from a distribution to $\mathcal{U}$, given that the norm we employ is (3.6).

## 4.2 A framework for distance estimates

We now restrict our attention to kernels of the form (3.7); for the sake of simplicity, we will assume the index set $\mathcal{A}$ in (3.7) is the set of nonnegative integers. For such kernels, we can use the framework below to help estimate the distance from a distribution in $H_{-s}$ to $\mathcal{U}$.

**Proposition 4.2** [1, Proposition 3.6] *Let $M$ be a positive integer, let $s > 0$, and let $\kappa$ have the form (3.7) with $a(k) > 0$ for all $k \geq 0$. If there are coefficients $c_1, \ldots, c_N$ such that, for $k = 0, \ldots, M$, $(w - \sum_{j=1}^{N} c_j u_j, \bar{F}_k) = 0$, and if there is a sequence $b(k) > 0$, $k = M+1, M+2, \ldots$, for which $|(w - \sum_{j=1}^{N} c_j u_j, \bar{F}_k)|^2 \leq b(k)$ when $k \geq M+1$, then*

$$\operatorname{dist}(w, \mathcal{U}) \leq \|w - \sum_{j=1}^{N} c_j u_j\| \leq \left( \sum_{k=M+1}^{\infty} a(k) b(k) \right)^{1/2}. \tag{4.3}$$

In addition to this proposition, we will use another result that connects the norm (3.6) with more familiar Sobolev norms. This result will apply to kernels of the form (3.7) for which the $a(k)$'s do not decay too rapidly. We remark that the two results are independent of one another.

**Proposition 4.3** [1, Proposition 3.5] *Let $s$, $\kappa$, and the $a(k)$'s be as in Proposition 4.2. If there is a $t > s$ for which*

$$a(k) \geq c^{-1}(1 + \lambda_k)^{-(s+t)/2} \tag{4.4}$$

*for all $k$ and some constant $c$, then every $f \in H_t(M^m)$ can be written as $f = \kappa \star v$ for some $v$ in $H_{-s}(M^m)$. Moreover,*

$$\|v\|_{-s} \leq c\|f\|_t \quad \text{and} \quad \|v\| \leq \|\kappa\|_{2s}^{1/2} c\|f\|_t. \tag{4.5}$$

We now turn to examples in which the framework established here is applied.

# 5 Applications

## 5.1 Tori

We will explain how our framework is used in a simple case: interpolating data at equally spaced points on the unit circle, $T^1$. In that case, we have that the distributions are $u_j = \delta_{2\pi j/N}$, where $j = 0, \ldots, N-1$. The eigenfunctions are of course

$$F_k(\phi) = \frac{e^{ik\theta}}{\sqrt{2\pi}}.$$

For our kernel we will use

$$\kappa(\phi, \phi') := \sum_{k=-\infty}^{\infty} a(k) e^{ik(\phi-\phi')},$$

which is of the form (3.7). We assume that we are given $w \in H_{-s}(T^1)$, and that the $a(k)$'s decay sufficiently fast for $P$ to be in $H_{2s}$. We will take $M = N-1$.

To apply Proposition 4.2 to this case, we must find coefficients $c_j$, $j = 0, \ldots, N-1$, such that $(w - \sum_{j=0}^{N-1} c_j u_j, e^{-ik\phi}/\sqrt{2\pi}) = 0$. In terms of the (usual) Fourier coefficients for $w$, we have

$$\widehat{w}(k) = \underbrace{\sum_{j=0}^{N-1} c_j e^{-2\pi i k j/N}}_{\widehat{c}_k},$$

where $k$ runs from $-[N/2]+1$ to $[(N-1)/2]$ and $\widehat{c}_k$ is the discrete Fourier transform (DFT) of the $c_j$'s. Inverting the DFT yields the $c_j$'s. It is easy to obtain $b(k)$'s so that $|(w - \sum_{j=0}^{N-1} c_j u_j, e^{-ik\phi}/\sqrt{2\pi})| \le b(k)$ for other values of $k$. The estimate on $\operatorname{dist}(w, \mathcal{U})$ from Proposition 4.2 is

$$\operatorname{dist}(w, \mathcal{U}) \le 2\|w\|_{-s} \left( \sum_{k \notin \mathcal{I}_N} (1+k^2)^s a_k \right)^{1/2},$$

where $\mathcal{I}_N := [-[N/2], [(N-1)/2]] \cap Z$. Using the inequality above in conjuction with (4.2) yields

$$|(\bar{w}, f - \tilde{f})| \le 4\|v\|_{-s}\|w\|_{-s} \left( \sum_{k \notin \mathcal{I}_N} (1+k^2)^s a_k \right).$$

If the coefficients $a(k)$ decay fast enough for the series to converge, but obey the condition in Proposition 4.3, then $\|v\|_{-s} \le c\|f\|_t$ and hence

$$|(\bar{w}, f - \tilde{f})| \le 4c\|f\|_t\|w\|_{-s} \left( \sum_{k \notin \mathcal{I}_N} (1+k^2)^s a_k \right).$$

This final bound is useful if one has *a priori* information about the 'energy' of the function $f$ generating the data. These results extend to $u_j$'s that are more general than point evaluations, include data that are only quasi-uniformly distributed, and carry over to some extent to $T^m$. See [1, §5].

## 5.2 The 2-sphere

In the interpolation problem that we wish to deal with here, we will use the physicist's convention of taking $\phi$ to be the azimuthal angle and $\theta$ the angle measured off the $z$-axis. The distributions will be point evaluations. Let $\Lambda$ be a fixed positive integer, and let $p_{j,k} \in S^2$ have coordinates $(\theta_j, \phi_k)$, where $\theta_j = \frac{\pi j}{2\Lambda}$, $\phi_k = \frac{\pi k}{\Lambda}$, and $j, k = 0, \ldots, 2\Lambda - 1$. We then take $u_{j,k} = \delta_{p_{j,k}}$.

The eigenfunctions for the Laplace–Beltrami operator on the 2-sphere are the spherical harmonics [7], $Y_{\ell,m}$, where $\ell = 0, 1, \ldots$ and $m = -\ell, \ldots, \ell$. We will use

$$\kappa(p,q) := \sum_{\ell=0}^{\infty} \sum_{m=-\ell}^{\ell} a(\ell,m) Y_{\ell,m}(p) \overline{Y_{\ell,m}(q)}.$$

Again these are of the form (3.7). As before, we assume that the $a(\ell,m)$'s decay quickly enough for $\kappa$ to be in $H_{2s}(S^2 \times S^2)$.

If $\Lambda$ is a power of 2, one can use Proposition 4.2 and a spherical version of the FFT [2] to get bounds on $\mathrm{dist}(w, \mathcal{U})$ [1, Theorem 6.6] similar to those given above in §5.1. We can use these bounds in (4.2) to obtain [1, Corollary 6.7]

$$|(\bar{w}, f - \tilde{f})| \le \Sigma(\Lambda)^2 \|v\|_{-s} \|w\|_{-s} \left( \sum_{\ell=\Lambda}^{\infty} \sum_{m=-\ell}^{\ell} (1 + \ell(\ell+1))^s a(\ell,m) \right)$$

where $\Sigma(\Lambda) := 1 + \frac{1}{\sqrt{\pi}} \Lambda^{3/2} \log_2(16\Lambda)$. Of course, if the $a(\ell,m)$'s satisfy (4.4), then by Proposition 4.3 we may replace $\|v\|_{-s}$ by $c\|f\|_s$. For further details, proofs, and precise statements of results, see [1, §6].

# References


[1] Dyn, N., F.J. Narcowich and J.D. Ward (1996) "Variational principles and Sobolev-type estimates for generalized interpolation on a Riemannian manifold", CAT Report # 371, Texas A&M University.

[2] Driscoll, J.R. and D.M. Healy (1994) "Computing Fourier transforms and convolutions for the 2-sphere", *Adv. in Appl. Math.* **15**, 202–250.

[3] Freeden, W. (1981) "On spherical spline interpolation and approximation", *Math. Meth. in the Appl. Sci.* **3**, 551–575.

[4] Madych, W.R. and S.A. Nelson (1988) "Multivariate interpolation and conditionally positive definite functions", *Approx. Theory and its Applications* **4**, 77–79.

[5] Madych, W.R. and S.A. Nelson (1990) "Multivariate interpolation and conditionally positive definite functions II", *Math. Comp.* **54**, 211–230.

[6] Menegatto, V.A. (1997) "Strictly positive definite kernels on the Hilbert sphere", *Applicable Analysis,* to appear.

[7] Müller, C. (1996) *Spherical Harmonics,* Springer-Verlag (Berlin).

[8] Narcowich, F.J. (1995) "Generalized Hermite interpolation and positive definite kernels on a Riemannian manifold", *J. Math. Anal. Applic.* **190**, 165–193.

[9] Narcowich, F.J. and J.D. Ward (1996) "Nonstationary wavelets on the $m$-sphere for scattered data", *Appl. Comp. Harm. Anal.* **3**, 324–336.

[10] Powell, M.J.D. (1990) "The theory of radial basis approximation in 1990", in *Wavelets, Subdivision and Radial Functions* (W. Light, ed.), Oxford University Press (Oxford).

[11] Ron, A. and X. Sun (1996) "Strictly positive definite functions on spheres", *Math. Comp.* **65**, 1513–1530.

[12] Schoenberg, I.J. (1942) "Positive definite functions on spheres", *Duke Math. J.* **9**, 96–108.

[13] Stewart, J. (1976) "Positive definite functions and generalizations, an historical survey", *Rocky Mountain J. Maths* **6**, 409–434.

[14] Wahba, G. (1981) "Spline interpolation and smoothing on the sphere", *SIAM J. Sci. Stat. Comput.* **2**, 5–16.

[15] Wahba, G. (1984) "Surface fitting with scattered noisy data on Euclidean $d$-space and on the sphere", *Rocky Mountain J. Maths* **14**, 281–299.

[16] Wendland, H. (1996) "Error estimates for interpolation by compactly supported radial basis functions of minimal degree", preprint, Institut für Numerische und Angewandte Mathematik, Universität Göttingen.

# Dense Factors of Sparse Matrices

*Roger Fletcher*

Department of Mathematics and Computer Science, University of Dundee, Dundee DD1 4HN, Scotland, UK.

**Abstract**

The method of Implicit LU factors for factorizing a nonsingular matrix $A$, and hence solving systems of equations, is described. It is shown how the factors are related to the regular LU factors computed by Gaussian Elimination. A backward error analysis is given and discussed. Implicit LU factors are shown to be advantageous when a copy of $A$ is kept for other purposes, and particularly so when $A$ is a sparse matrix stored in compact form. It is shown how the method can be implemented efficiently when there are many unit columns in $A$.

Techniques for updating Implicit LU factors are described when one column in $A$ is replaced by a new column, with applications to linear programming and related calculations. Ideas in the Fletcher–Matthews method for updating regular LU factors are adapted to update implicit factors, and the resulting method is seen to be more efficient.

Other possibilities for implicit factors are also described which use more storage but are potentially more stable. Some comparisons on a range of linear programming problems are presented and discussed.

## 1 Introduction

The research in this paper arises in the context of Simplex-like methods for Linear Programming (LP) and related calculations. Such methods involve a nonsingular $n \times n$ matrix $A$, and systems of linear equations involving both $A$ and $A^T$ are required to be solved. On each iteration of the Simplex method a column is removed from $A$ and a new column is added (the Simplex update). The columns that arise in the LP problem, and hence in $A$, are usually sparse and are stored in compact format.

To solve the very largest LP problems, say $10^5$ variables or more, very specialized product form methods have been devised (see for example Suhl [21]). Development of codes for these methods is very intricate and it is common for some man-years of effort to be required. Also there are some concerns about numerical stability.

For smaller LP problems, say 1000 variables or less, it becomes more attractive to compute dense factors, avoiding problems with fill-in and product forms. This gives a simple code which is readily understood and maintained, and hopefully permits better numerical stability. It is nonetheless desirable to take advantage of the fact that $A$ is stored in compact form.

An obvious candidate would be to use regular LU or QR factors which are readily updated and have exhibited excellent numerical stability in practice. This paper argues that it can be much better to use so called Implicit LU factors with significantly less storage and operations counts, yet with good numerical stability.

Ideas of implicit factorizations go back many years and I am grateful to Michele Benzi of CERFACS for information on some of the early papers. Different types of implicit factors are possible, and there are relationships with iterative methods such as conjugate gradients and projection methods. More recently, those familiar with null-space methods will also see a connection. To give an idea of the frequency with which these ideas have been discovered and rediscovered, some the earlier papers are itemized in the following table.

| | | |
|---|---|---|
| Fox, Huskey and Wilkinson | [9] | 1948 |
| Hestenes and Stiefel | [12] | 1952 |
| Purcell | [16] | 1953 |
| Householder | [13] | 1955 |
| Pietrzykowski | [15] | 1960 |
| Faddeev and Faddeeva | [7] | 1963 |
| Stewart | [19] | 1973 |
| Enderson and Wassyng | [6] | 1978 |
| Sloboda | [18] | 1978 |
| Wassyng | [23] | 1982 |
| Abaffy, Broyden and Spedicato | [1] | 1984 |
| Hegedüs | [11] | 1986 |

Most of the papers develop the method as an iterative method based on conjugacy and orthogonality properties, although Purcell does recognise the method as being a competitor of the Crout decomposition. The term Implicit LU factors is first used in the paper of Abaffy, Broyden and Spedicato and they give more references to previous related work. Abaffy, Broyden and Spedicato regard Implicit LU factors as a special case of a so-called ABS method, and there are many more recent contributions by Spedicato and his co-workers. Other recent researchers to use the idea are Tuma [22] and Benzi and Meyer [3], [4].

Despite these many references, the ideas involved are not well known in the numerical analysis community, perhaps because they are not aired in standard texts, and perhaps because the diversity of previous papers has not established a common style for presentation of the algorithms. Also, algorithm (2.15) given below does not appear to be referenced elsewhere, so the potential of Implicit LU factors for facilitating solves with both $A$ and $A^T$ may not have been recognised.

This paper makes some other contributions which are thought to be new, including a backward error analysis (Section 3), application to block systems (Section 4), and the derivation of updating techniques (Section 5). A comparison of other types of implicit factors (which are special cases of ABS methods) is described and discussed in Section 7, and provides evidence that the efficiency of Implicit LU factors is obtained with only marginal loss of numerical stability. Hopefully this paper will also lead to an enhanced awareness of the existence of implicit factors, leading to their increased use in other situations.

# 2 Implicit LU factors

In this section the idea of Implicit LU factors of an $n \times n$ nonsingular matrix $A$ is explained and its relationship to the regular LU factors that may be computed by Gaussian Elimination (GE) is described. It is also shown how Implicit LU factors may be used to solve well-determined systems of linear equations.

First the computation and use of regular LU factors by GE is reviewed, using a compact matrix notation. The details of the computation are well-known (e.g. Wilkinson [24]) and can be expressed in the form

$$A^{(k+1)} = M^{(k)}A^{(k)} \qquad k = 1, 2, \ldots, n-1 \tag{2.1}$$

where $A^{(1)} = A$. $M^{(k)}$ differs from a unit matrix only in that $(M^{(k)})_{jk} = -m_{jk}$ for $j = k+1, \ldots, n$, where

$$m_{jk} = a_{jk}^{(k)}/a_{kk}^{(k)} \qquad j = k+1, \ldots, n \tag{2.2}$$

are the so-called *multipliers* at stage $k$. The operation in (2.1) eliminates the subdiagonal elements in column $k$ of $A^{(k)}$ so that the generic form of $M^{(k)}$ and $A^{(k)}$ is typified for $n = 6$ and $k = 3$ by

$$M^{(k)} = \begin{bmatrix} 1 & & & & & \\ & 1 & & & & \\ & & 1 & & & \\ & & -m_{43} & 1 & & \\ & & -m_{53} & & 1 & \\ & & -m_{63} & & & 1 \end{bmatrix}$$

and

$$A^{(k)} = \begin{bmatrix} a_{11}^{(k)} & a_{12}^{(k)} & a_{13}^{(k)} & a_{14}^{(k)} & a_{15}^{(k)} & a_{16}^{(k)} \\ & a_{22}^{(k)} & a_{23}^{(k)} & a_{24}^{(k)} & a_{25}^{(k)} & a_{26}^{(k)} \\ & & a_{33}^{(k)} & a_{34}^{(k)} & a_{35}^{(k)} & a_{36}^{(k)} \\ & & a_{43}^{(k)} & a_{44}^{(k)} & a_{45}^{(k)} & a_{46}^{(k)} \\ & & a_{53}^{(k)} & a_{54}^{(k)} & a_{55}^{(k)} & a_{56}^{(k)} \\ & & a_{63}^{(k)} & a_{64}^{(k)} & a_{65}^{(k)} & a_{66}^{(k)} \end{bmatrix}.$$

It is noted that the inverse of $M^{(k)}$ is obtained simply by changing the signs of the subdiagonal elements of $M^{(k)}$. Then it is readily verified that $A = LU$ where

$$L = M^{(1)^{-1}} M^{(2)^{-1}} \ldots M^{(n-1)^{-1}} \quad \text{and} \quad U = A^{(n)}. \tag{2.3}$$

Equivalently $L$ is obtained simply by adding the multipliers $m_{ij}$ into the corresponding subdiagonal elements of a unit matrix, for all $i > j$. A row interchange may be carried out at the start of each stage (partial pivoting) to ensure that the pivot $a_{kk}^{(k)}$ is non-zero, and commonly the pivot with largest modulus is chosen. This ensures that the elements of $L$ are bounded in modulus by 1. The operations count for the computation of $L$ and $U$ is $\frac{1}{3}n^3 + O(n^2)$ multiplications, with a similar number of additions.

To solve a system $A\mathbf{x} = \mathbf{b}$, the systems

$$L\mathbf{y} = \mathbf{b} \qquad \text{and} \qquad U\mathbf{x} = \mathbf{y} \tag{2.4}$$

are solved in turn. These are triangular systems and may be solved by forward and backward substitution respectively, each substition requiring $\frac{1}{2}n^2 + O(n)$ multiplications. Likewise to solve $A^T\mathbf{x} = \mathbf{b}$, the systems

$$U^T\mathbf{y} = \mathbf{b} \qquad \text{and} \qquad L^T\mathbf{x} = \mathbf{y} \tag{2.5}$$

are solved in a similar way.

An algorithm to determine one particular type of Implicit LU factors, which we refer to as LIU factors, may now be described. It computes a lower triangular matrix $\mathbb{L}$ and a diagonal matrix $D = \text{diag}(d_1, d_2, \ldots, d_n)$. It is convenient to introduce an intermediate lower triangular matrix $\mathbb{L}^{(k)}$, the subdiagonal elements of which are zero in columns $k, \ldots, n-1$. For $n = 6$ and $k = 3$ this would be typified by

$$\mathbb{L}^{(k)} = \begin{bmatrix} 1 & & & & & \\ l_{21}^{(k)} & 1 & & & & \\ l_{31}^{(k)} & l_{32}^{(k)} & 1 & & & \\ l_{41}^{(k)} & l_{42}^{(k)} & & 1 & & \\ l_{51}^{(k)} & l_{52}^{(k)} & & & 1 & \\ l_{61}^{(k)} & l_{62}^{(k)} & & & & 1 \end{bmatrix}. \tag{2.6}$$

Initially $\mathbb{L}^{(1)} = I$ and the algorithm is

$$\begin{array}{l} \texttt{for } k = 1, 2, \ldots, n-1 \\ \quad d_k = \mathbf{l}_k^{(k)T}\mathbf{a}_k \\ \quad \texttt{for } j = 1, \ldots, k \\ \qquad \mathbf{l}_j^{(k+1)} = \mathbf{l}_j^{(k)} \\ \quad \texttt{end} \\ \quad \texttt{for } j = k+1, \ldots, n \\ \qquad r_{jk} = \mathbf{l}_j^{(k)T}\mathbf{a}_k / d_k \\ \qquad \mathbf{l}_j^{(k+1)} = \mathbf{l}_j^{(k)} - r_{jk}\mathbf{l}_k^{(k)} \\ \quad \texttt{end} \\ \texttt{end} \end{array} \tag{2.7}$$

where $\mathbf{l}_j^{(k)T}$ denotes row $j$ of $\mathbb{L}^{(k)}$ and $\mathbf{a}_k$ denotes column $k$ of $A$. Finally $\mathbb{L} = \mathbb{L}^{(n)}$ and $d_n = \mathbf{l}_n^{(n)T}\mathbf{a}_n$.

The relationship of this algorithm to GE is now considered. The proposition that

$$\mathbb{L}^{(k)}A = A^{(k)} \tag{2.8}$$

is established inductively, where $A^{(k)}$ is the matrix introduced in (2.1). Clearly (2.8) is true for $k = 1$. If (2.8) is true for some $k \geq 1$, it follows that

$$a_{jk}^{(k)} = \mathbf{l}_j^{(k)T}\mathbf{a}_k \qquad j = k, k+1, \ldots, n. \tag{2.9}$$

Hence the ratios

$$r_{jk} = \frac{\mathbf{l}_j^{(k)T}\mathbf{a}_k}{d_k} = \frac{a_{jk}^{(k)}}{a_{kk}^{(k)}} = m_{jk} \qquad j = k+1, \ldots, n \tag{2.10}$$

are just the multipliers in GE. Thus the LIU calculation in the inner loops of (2.7) can be expressed as

$$\mathbb{L}^{(k+1)} = M^{(k)}\mathbb{L}^{(k)}. \tag{2.11}$$

It follows that

$$A^{(k+1)} = M^{(k)}A^{(k)} = M^{(k)}\mathbb{L}^{(k)}A = \mathbb{L}^{(k+1)}A.$$

This is (2.8) with $k$ replaced by $k+1$, which establishes (2.8) for all $k > 1$.

In the case $k = n$ it is deduced from (2.8) and (2.3) that

$$\mathbb{L}A = U \tag{2.12}$$

and it is seen that

- $\mathbb{L}$ for the LIU factors is the *inverse* of $L$ for the regular LU factors
- $D$ for the LIU factors is the *diagonal* of $U$ for the regular LU factors

(since $U = A^{(n)}$ and $D = \text{diag}(a_1^{(1)}, a_2^{(2)}, \ldots, a_n^{(n)})$). The rest of the matrix $U$ is not explicitly available in the LIU approach but is given implicitly by the equations

$$u_{ij} = a_{ij}^{(n)} = \mathbf{l}_i^T\mathbf{a}_j \qquad i < j \tag{2.13}$$

where $\mathbf{l}_i^T$ denotes row $i$ of $\mathbb{L}$. This explains the use of the term LIU (for L-Implicit-U) factors. It is also readily possible to calculate (Implicit-L)-U factors of $A$ by carrying out *column* operations on $A$, eliminating elements above the diagonal. Indeed, this form seems to be the one that is used in all previous work. Of course this is equivalent to obtaining the LIU factors of $A^T$. However, the form used in this paper has the advantage of being more closely related to the familiar form of GE in which $L$ is unit triangular, and row operations and row pivoting are used. This discussion also brings out the lack of symmetry in LIU factors and it can be seen that the algorithm does not compete with Choleski factors when $A$ is symmetric and positive definite (except possibly for parallel computation, see Benzi and Meyer [4]).

For dense $A$, it is possible for $\mathbb{L}$ and $D$ to overwrite the lower triangle of $A$ as the calculation proceeds, in which case the LIU algorithm (2.7) requires $\frac{1}{3}n^3+O(n^2)$ multiplications, as for GE. Also it will be seen below that only the strict upper triangle of $A$ is required when solving systems with LIU factors. Likewise, only the strict upper triangle of $A$ is required to recover $U$ from (2.13). Thus the lower triangle of $A$ can be replaced by $\mathbb{L}$, $D$ and its strict upper triangle, without any essential loss of information (ignoring round-off). Of course for sparse $A$ this property is unlikely to be advantageous.

Partial pivoting in the LIU algorithm is possible in a similar way to GE. To do this, the elements $a_{jk}^{(k)} = \mathbf{l}_j^{(k)T}\mathbf{a}_k$ are assembled at the start of stage $k$ of (2.7), and a row interchange is made which brings the element with maximum modulus onto the diagonal, to become $d_k$. This gives rise to factors

$$\mathbb{L}PA = U \tag{2.14}$$

and is exactly equivalent to partial pivoting in GE. However, in contrast to GE, complete pivoting is not practicable in the LIU algorithm because $A^{(k)}$ is not stored during the calculation of $\mathbb{L}$ and $D$, and computation of the relevant elements of $A^{(k)}$ would be too expensive. To simplify the subsequent presentation, it is assumed that the row permutations have been incorporated into $A$ so that (2.14) can be written as (2.12).

The use of LIU factors in the solution of linear systems is now considered. Equations (2.4) and (2.5) cannot be used directly as $U$ is not explicitly available. To solve a system $A\mathbf{x} = \mathbf{b}$, it is written as $\mathbb{L}A\mathbf{x} = \mathbb{L}\mathbf{b}$ and hence as $U\mathbf{x} = \mathbb{L}\mathbf{b}$ using (2.12). Thus, using (2.13),

$$\begin{aligned} x_i &= \left(\mathbf{l}_i^T\mathbf{b} - \sum_{j=i+1}^{n} u_{ij}x_j\right) \Big/ u_{ii} \\ &= \left(\mathbf{l}_i^T\mathbf{b} - \sum_{j=i+1}^{n} \mathbf{l}_i^T\mathbf{a}_j x_j\right) \Big/ d_i \\ &= \mathbf{l}_i^T\left(\mathbf{b} - \sum_{j=i+1}^{n} \mathbf{a}_j x_j\right) \Big/ d_i. \end{aligned}$$

The bracketed quantity is seen to be the partial sum in the calculation of the residual vector $\mathbf{b} - A\mathbf{x}$, giving rise to the algorithm

$$\begin{aligned} &\mathbf{b}^{(n)} = \mathbf{b} \\ &\texttt{for}\ \ i = n, n-1, \ldots, 1 \\ &\quad x_i = \mathbf{l}_i^T\mathbf{b}^{(i)}/d_i \\ &\quad \mathbf{b}^{(i-1)} = \mathbf{b}^{(i)} - \mathbf{a}_i x_i \\ &\textbf{end}. \end{aligned} \tag{2.15}$$

This algorithm may be new and is certainly not well known. Because of the structure of $\mathbf{l}_i$, only a column of the strict upper triangle of $A$ need be used in the update of $\mathbf{b}^{(i+1)}$. When $A$ is a dense matrix, this enables the calculation to be completed in $n^2 + O(n)$ multiplications, the same as for (2.4).

To solve a system $A^T\mathbf{x} = \mathbf{b}$, the vector $\mathbf{y}$ defined by $\mathbf{x} = \mathbb{L}^T\mathbf{y}$ is introduced, as in (2.5). Then $A^T\mathbf{x} = \mathbf{b}$ can be written as $A^T\mathbb{L}^T\mathbf{y} = \mathbf{b}$ and hence as $U^T\mathbf{y} = \mathbf{b}$ using (2.12). Thus, using (2.13),

$$\begin{aligned} y_i &= \left(b_i - \sum_{j=1}^{i-1} u_{ji}y_j\right) \Big/ u_{ii} \\ &= \left(b_i - \sum_{j=1}^{i-1} \mathbf{a}_i^T\mathbf{l}_j y_j\right) \Big/ d_i \\ &= \left(b_i - \mathbf{a}_i^T \sum_{j=1}^{i-1} \mathbf{l}_j y_j\right) \Big/ d_i \end{aligned}$$

In this case, the summation is seen to involve the partial sum in the calculation of the solution vector $\mathbf{x} = \mathbb{L}^T\mathbf{y} = \sum_{i=1}^n \mathbf{l}_i y_i$, giving rise to the algorithm

$$\begin{aligned} &\mathbf{x}^{(0)} = \mathbf{0} \\ &\texttt{for}\ \ i = 1, 2, \ldots, n \\ &\quad y_i = (b_i - \mathbf{a}_i^T\mathbf{x}^{(i-1)})/d_i \\ &\quad \mathbf{x}^{(i)} = \mathbf{x}^{(i-1)} + \mathbf{l}_i y_i \\ &\textbf{end} \\ &\mathbf{x} = \mathbf{x}^{(n)}. \end{aligned} \tag{2.16}$$

In this algorithm, $\mathbf{x}^{(i-1)}$ has the same structure as $\mathbf{l}_{i-1}$, so only column $i$ of the strict upper triangle of $A$ need be used in the calculation of $y_i$. As before, when $A$ is a dense matrix, this enables the calculation to be completed in $n^2 + O(n)$ multiplications. In fact the algorithm may be initialized with $\mathbf{x}^{(0)}$ other than $\mathbf{0}$, and can be interpreted as an iterative projection method based on the orthogonality properties $\mathbf{l}_i^T\mathbf{a}_j = 0, \quad i > j$ that derive from (2.12). Many references have developed the LIU algorithm from this point of view.

# 3 An error analysis of LIU factors

The floating point error analysis of Gaussian Elimination and triangular substitution (Wilkinson [24]) has provided a thorough understanding of the numerical stability properties of these algorithms. The computed factors in GE are shown to be the exact factors $LU = A + E$ of a perturbed system, where $E$ is referred to as the backward error. $E$ can be bounded by an expression of the form $3\rho n\varepsilon$ in which $\rho$ measures the growth in $A^{(k)}$ during the elimination and $\varepsilon$ is the relative precision of the arithmetic. Thus the purpose of the pivoting process is seen to be one of inhibiting growth during the elimination. For a triangular system the computed solution is also the exact solution of a perturbed system in which the perturbation is bounded by a modest multiple of $\varepsilon$. These bounds may be combined to give a total backward error bound for solving $A\mathbf{x} = \mathbf{b}$ by GE of the form $\rho\phi(n)\varepsilon$ in which $\phi(n)$ is a modest quadratic in $n$ (Stewart [20], Theorem 5.3). This bound usually overstates the dependence on $n$ which is unlikely to be a dominant factor. If partial pivoting is used, exponential growth is possible, but in practice this is very rare. Ill-conditioned systems usually do not lead to growth in the backward error, and pose no particular difficulty for the algorithm.

It is with this background in mind that an error analysis of the LIU factorization and solution algorithms has been attempted. It is shown that a backward error expression for the factorization algorithm (2.7) may be obtained. Bounds on the backward error, akin to those for GE, are derived. The main feature is seen to be the need to avoid growth in the matrix $\mathbb{L}$. It is argued that, with partial pivoting, growth is no more likely with LIU factors than with LU factors. Hence it is concluded that there is little to choose between the two algorithms on the grounds of numerical stability. It has not proved possible to obtain a satisfactory backward error bound for the solution algorithms (2.15) and (2.16). The reasons for this are explained. It is argued that this is unlikely to cause significant deterioration in accuracy unless the magnitude of the solution $\mathbf{x}$ is large.

The main result for algorithm (2.7) in floating point arithmetic of relative precision $\varepsilon$ is:

**Theorem 1** *There exists a lower triangular matrix $E$ such that the computed LIU factors $\mathbb{L}$ and $D$ are the exact factors of a matrix $A+E$, that is*

$$\mathbb{L}(A+E) = U, \tag{3.1}$$

*where $U$ is upper triangular, $D = \operatorname{diag}(u_{11}, u_{22}, \ldots, u_{nn})$ and $u_{jk}$ is given implicitly by $\mathbf{l}_j^T \mathbf{a}_k$ for all $j < k$. $E$ may be bounded by*

$$|e_{jk}| \le (\tfrac{3}{2}n^2 + O(n))\varepsilon\gamma_{jk} + O(\varepsilon^2) \tag{3.2}$$

*where*

$$\gamma_{jk} = \max_{p \ge k}\{\max_i |l_{ji}^{(p)} a_{ik}|\}. \tag{3.3}$$

**Proof** The errors that arise in processing a column $\mathbf{a}_k$ of $A$ are considered. This column first arises at stage $k$ of (2.7), and is used to calculate

$$a_{jk}^{(k)} = \mathrm{fl}(\mathbf{l}_j^{(k)T}\mathbf{a}_k)) \qquad j = k, \ldots, n$$

($\mathrm{fl}(\cdot)$ indicates the outcome of a calculation in floating point arithmetic), followed by

$$m_{jk} = \mathrm{fl}(a_{jk}^{(k)}/a_{kk}^{(k)}) \qquad j = k+1, \ldots, n.$$

A vector $\delta^{(k)}$ to account for these errors may be defined by

$$\begin{aligned} &\delta_j^{(k)} = 0, \quad j < k, \qquad \delta_k^{(k)} = a_{kk}^{(k)} - \mathbf{l}_k^{(k)T}\mathbf{a}_k, \\ &\delta_j^{(k)} = m_{jk}a_{kk}^{(k)} - \mathbf{l}_j^{(k)T}\mathbf{a}_k, \quad j > k. \end{aligned} \tag{3.4}$$

It then follows by virtue of the structure of $\mathbb{L}^{(k)}$ (see (2.6)) that

$$\mathbf{l}_k^{(k)T}(\mathbf{a}_k + \delta^{(k)}) = a_{kk}^{(k)} \qquad \text{and} \qquad \mathbf{l}_j^{(k)T}(\mathbf{a}_k + \delta^{(k)}) = m_{jk}a_{kk}^{(k)} \quad j > k,$$

so that if the calculation of $a_{jk}^{(k)}$ and $m_{jk}$ is carried out on $\mathbf{a}_k + \delta^{(k)}$ in exact arithmetic, then $a_{jk}^{(k+1)} = 0$ is obtained. Thus in exact arithmetic we can write

$$M^{(k)}\mathbb{L}^{(k)}(\mathbf{a}_k + \delta^{(k)}) = \mathbf{u}_k \tag{3.5}$$

where $u_{kk} = a_{kk}^{(k)} = d_k$, $u_{jk} = 0$ for $j > k$, and $u_{jk}$ is implicitly defined by $\mathbf{l}_j^{(k)T}\mathbf{a}_k$ for $j < k$.

To obtain bounds on $\delta^{(k)}$, standard error analysis techniques are used. Thus we write $m_{jk} = (1 + f_{jk})a_{jk}^{(k)}/a_{kk}^{(k)}$ where $|f_{jk}| \leq \varepsilon$ so that

$$|m_{jk}a_{kk}^{(k)} - a_{jk}^{(k)}| \leq \varepsilon |a_{jk}^{(k)}| \qquad j > k.$$

Similarly we may express

$$\begin{aligned} a_{jk}^{(k)} &= \mathrm{fl}(\mathbf{l}_j^{(k)T}\mathbf{a}_k)) \\ &= (\ldots((a_{jk} + l_{j1}a_{1k}(1 + e_1^{jk}))(1 + f_1^{jk}) + l_{j2}a_{2k}(1 + e_2^{jk}))(1 + f_2^{jk}) + \ldots \\ &\quad + l_{j,k-1}a_{k-1,k}(1 + e_{k-1}^{jk}))(1 + f_{k-1}^{jk}) \end{aligned}$$

where $e_i^{jk}$ and $f_i^{jk}$ are relative errors bounded by in modulus by $\varepsilon$. Collecting the errors in each term, and using $a_{jk} = l_{jj}^{(k)}a_{jk}$ and (3.4), gives

$$\begin{aligned} |\delta_k^{(k)}| &\leq k^2\varepsilon \max_i |l_{ki}^{(k)}a_{ik}| + O(\varepsilon^2) \\ |\delta_j^{(k)}| &\leq k^2\varepsilon \max_i |l_{ji}^{(k)}a_{ik}| + \varepsilon|a_{jk}^{(k)}| + O(\varepsilon^2) \\ &\leq (k^2 + k)\varepsilon \max_i |l_{ji}^{(k)}a_{ik}| + O(\varepsilon^2) \qquad j > k. \end{aligned} \tag{3.6}$$

Further errors are incurred when $\mathbb{L}^{(p+1)} = \mathrm{fl}(M^{(p)}\mathbb{L}^{(p)})$ is calculated, both for $p = k$ and $p > k$, which can be accounted for by making a further perturbation of $\mathbf{a}_k$. The detailed effect of errors in calculating $\mathbb{L}^{(p+1)}$ is now considered. Since $l_{ji}^{(p+1)} = l_{ji}^{(p)}$ for $j \leq p$ and all $i$, there is no error in this case. Likewise for $j > p$ and $i > p$. Moreover $l_{jp}^{(p+1)} = -m_{jp}$ for $j > p$ so there is also no error in this case. Rounding errors only arise for $j > p$ and $i < p$, defined by

$$\begin{aligned} l_{ji}^{(p+1)} &= (l_{ji}^{(p)} - m_{jp}l_{pi}^{(p)}(1 + g_{ji}^{(p)}))(1 + h_{ji}^{(p)}) \\ &= l_{ji}^{(p)} - m_{jp}l_{pi}^{(p)} + h_{ji}^{(p)}l_{ji}^{(p+1)} + g_{ji}^{(p)}(l_{ji}^{(p+1)} - l_{ji}^{(p)}) + O(\varepsilon^2) \end{aligned}$$

where $g_{ji}^{(p)}$ and $h_{ji}^{(p)}$ are relative errors bounded in modulus by $\varepsilon$. This can be expressed in the form

$$\mathbb{L}^{(p+1)} = M^{(p)}\mathbb{L}^{(p)} + E^{(p)} \tag{3.7}$$

where

$$e_{ji}^{(p)} = h_{ji}^{(p)} l_{ji}^{(p+1)} + g_{ji}^{(p)} (l_{ji}^{(p+1)} - l_{ji}^{(p)}) + O(\varepsilon^2). \tag{3.8}$$

Note that $E^{(p)}$ has zeros throughout rows $1, \ldots, p$ and columns $p, \ldots, n$. It also follows that

$$\left(E^{(p)}\mathbf{a}_k\right)_j = \sum_{i=1}^{p-1} (h_{ji}^{(p)} l_{ji}^{(p+1)} + g_{ji}^{(p)} (l_{ji}^{(p+1)} - l_{ji}^{(p)})) a_{ik} + O(\varepsilon^2)$$

and we can deduce the bound

$$\left|\left(E^{(p)}\mathbf{a}_k\right)_j\right| \leq 3(p-1)\varepsilon \max_i \{\max(|l_{ji}^{(p+1)} a_{ik}|, |l_{ji}^{(p)} a_{ik}|)\} + O(\varepsilon^2) \tag{3.9}$$

for $j > p$ and $p \geq k$.

Equation (3.7) with $p = k$ can be substituted into (3.5) to give

$$(\mathbb{L}^{(k+1)} - E^{(k)})(\mathbf{a}_k + \delta^{(k)}) = \mathbf{u}_k$$

and hence

$$\mathbb{L}^{(k+1)}(\mathbf{a}_k - E^{(k)}\mathbf{a}_k + \delta^{(k)}) = \mathbf{u}_k$$

since $E^{(k)}\delta^{(k)} = \mathbf{0}$. Subsequent operations on $\mathbb{L}^{(p)}$ for $p > k$ can be accommodated in a similar way (note that $M^{(p)}\mathbf{u}_k = \mathbf{0}$) to give

$$\mathbb{L}\left(\mathbf{a}_k + \delta^{(k)} - \sum_{p=k}^{n-1} E^{(p)}\mathbf{a}_k\right) = \mathbf{u}_k. \tag{3.10}$$

The error terms in (3.10) can all be bounded using the growth factor (3.3) and the bounds in (3.6) and (3.9). The leading coefficient for the total error $E_{jk}$ is

$$k^2 + k + 3\sum_{p=k}^{n} p$$

which reduces to $\frac{3}{2}n^2 - \frac{1}{2}k^2$ and hence yields the result in (3.1). □

Some discussion on the bounds given by Theorem 1 is appropriate. The quantities $l_{ji}^{(p)} a_{ik}$, $i = 1, 2, \ldots, n$ in (3.3) are the individual terms in the scalar product $\mathbf{l}_j^{(p)} \mathbf{a}_k$ that defines $a_{jk}^{(p)}$. It is the factor $\max_p |a_{jk}^{(p)}|$ that features in the error bound for the computation of LU factors by GE. Thus these error bounds are closely related. The factor $\frac{3}{2}n^2$ in place of $3n$ arises in (3.3), essentially because the individual terms in the scalar product have been used. However the total bounds for solving $A\mathbf{x} = \mathbf{b}$ by GE (Stewart, [20]) also contain an $n^2$ factor. These factors of $n^2$ or $n$ are likely to overestimate the actual dependence of the error on $n$, and it is unlikely that the occurrence of an $n^2$ factor indicates a significantly worse dependence on $n$. However, the calculation of any term $l_{ji}^{(p)} a_{ik}$ does create an error $\sim \varepsilon l_{ji}^{(p)} a_{ik}$ and so the dependence of the error bound on the growth factor $\gamma_{jk}$ defined in (3.3) is likely to be realistic in practice. Note also that, as for the GE analysis, Theorem 1 does not make an assumption that partial pivoting has been used.

The message conveyed by (3.1) is that growth in the backward error is correlated with growth in any of the matrices $\mathbb{L}^{(p)}$ for $p = 1, \ldots, n$. If partial pivoting is used, then $|m_{jk}| \leq 1$ and the growth can be bounded by $2^{n-2}$. This is attained for the matrix

$$A = \begin{bmatrix} 1 & 0 & 0 & \cdots & 0 & 1 \\ -1 & 1 & 0 & \cdots & 0 & 1 \\ -1 & -1 & 1 & \cdots & 0 & 1 \\ & & & \ddots & & \\ -1 & -1 & -1 & \cdots & 1 & 1 \\ -1 & -1 & -1 & \cdots & -1 & 1 \end{bmatrix}.$$

This is the same matrix for which the growth bound on $U$ for GE is obtained. In practice serious growth is very rare, even if the bound on the multipliers is relaxed, as in some sparse matrix methods. These considerations, and the similarity of the backward error bounds, suggest that the numerical stability of the calculation of both LIU factors and LU factors is very satisfactory, and there is little to choose between the algorithms in this respect.

Another consequence of (3.1) is that

$$\mathbb{L}A = U + \mathbb{L}^{-1}E \tag{3.11}$$

where $\mathbb{L}^{-1}E$ is also lower triangular. Now $\mathbb{L}^{-1}$ is the matrix $L$ in the LU factors of $A$. If partial pivoting is used then the elements of $L$ are in $[-1, 1]$ and hence $\mathbb{L}^{-1}E$ has a similar sort of bound to that for $E$.

Turning to the error analysis of the solution algorithms (2.15) and (2.16) that use LIU factors, most of the errors that arise can be thrown onto the data, as in Wilkinson's [24] analysis of triangular substitution. However, the errors arising from the partial sums $\mathbf{b}^{(i)}$ (for (2.15)) and $\mathbf{x}^{(i)}$ (for (2.16)) are not obviously disposed of in this way. This may indicate a potential source of instability if significant growth in the partial sums occurs. For GE there is a good backward error bound for the entire process of solving $A\mathbf{x} = \mathbf{b}$ or $A^T\mathbf{x} = \mathbf{b}$. An important consequence is that if $\mathbf{x} \sim O(1)$ then the solution process gives an accurate residual error *that is unaffected by any ill-conditioning in A*. However, in this case the partial sums arising from (2.15) and (2.16) are unlikely to grow, in which case accurate residuals are also likely to be obtained. This outcome has been verified by some Matlab experiments on increasingly ill-conditioned matrices.

# 4 LIU factors for sparse matrices

LIU factors have some advantages when $A$ is a sparse matrix that is stored in compact form. Assuming that there is a low proportion of non-zero elements in $A$, the operations involving $A$ in (2.7), (2.15) and (2.16) have negligible cost. There is a disadvantage that $\mathbb{L}$ is not usually very sparse (if $A$ is block irreducible, $\mathbb{L}$ is generally fully dense), although Benzi and Meyer [4] have had some success with the use of drop tolerances in conjunction with Implicit LU factors. For very large sparse matrices, sparse LU factors are likely to provide the method of choice since both $L$ and $U$ often retain a significant amount of the sparsity in $A$. Nevertheless coding a sparse LU solver is a complex task because of the need to account for fill-in during the factorization. Updating such representations in algorithms such as the Simplex method for Linear Programming is also complex and there are some concerns in regard to numerical stability. If $A$ is sparse but not too large ($n$ is less than 1000, say) then it becomes attractive to consider the use of dense LIU factors. The storage required is $\frac{1}{2}n^2 + O(n)$ locations and the major cost is that of sdot or saxpy operations with rows of $\mathbb{L}$, which can be coded very efficiently. Moreover LIU factors are readily and stably updated when simplex column updates are made to $A$, as is demonstrated in Section 5.

In fact the situation can be even more favourable when $A$ has many unit columns, a situation that often arises in Linear Programming applications. In this case we may express

$$A = \begin{bmatrix} A_1 & 0 \\ A_2 & I \end{bmatrix}$$

without loss of generality, where $A_1$ is $m \times m$ nonsingular, and $I$ has dimension $n - m$. However, more is gained if we permute this matrix so that $A_2$ is in the upper triangle, and so is handled implicitly. Thus we consider the matrix

$$A = \begin{bmatrix} I & A_2 \\ 0 & A_1 \end{bmatrix} \tag{4.1}$$

for which the subdiagonal elements of $\mathbb{L}$ are zero in columns $1, \ldots, n - m$. A particularly nice feature which does not seem to have been observed previously, is that the solution algorithms do not need to partition $A$ into $A_1$ and $A_2$, but can work with columns of $A$ itself. It is readily verified that algorithm (2.15) for solving $A\mathbf{x} = \mathbf{b}$ simplifies to become

$$\begin{array}{l} \mathbf{b}^{(n)} = \mathbf{b} \\ \texttt{for } i = n, n-1, \ldots, n-m+1 \\ \quad x_i = \mathbf{l}_i^T \mathbf{b}^{(i)} / d_i \\ \quad \mathbf{b}^{(i-1)} = \mathbf{b}^{(i)} - \mathbf{a}_i x_i \\ \textbf{end} \\ \texttt{for } i = 1, \ldots, n-m \\ \quad x_i = b_i^{(n-m)} \\ \textbf{end.} \end{array} \tag{4.2}$$

Also the algorithm for solving $A^T\mathbf{x} = \mathbf{b}$ becomes

$$\begin{array}{l}
\texttt{for } i = 1, \ldots, n-m \\
\quad \mathbf{x}_i^{(n-m)} = b_i \\
\texttt{end} \\
\texttt{for } i = n-m+1, \ldots, n \\
\quad \mathbf{x}_i^{(n-m)} = 0 \\
\texttt{end} \\
\texttt{for } i = n-m+1, \ldots, n \\
\quad y_i = (b_i - \mathbf{a}_i^T \mathbf{x}^{(i-1)})/d_i \\
\quad \mathbf{x}^{(i)} = \mathbf{x}^{(i-1)} + \mathbf{l}_i y_i \\
\texttt{end} \\
\mathbf{x} = \mathbf{x}^{(n)}.
\end{array} \tag{4.3}$$

The dominant cost for each is $\sim \frac{1}{2}m^2$ multiplications, assuming that the operations with $\mathbf{a}_i$ have negligible cost.

# 5 Simplex updates of LIU factors

In the Simplex method for Linear Programming and related calculations, an important step is to update factors of $A$ efficiently when one column of $A$ is replaced by a new column. The ordering of the columns in $A$ is immaterial. An efficient and reasonably stable algorithm for updating $PA = LU$ factors is that of Fletcher and Matthews [8], and this section shows how the same ideas can be adapted for updating LIU factors. These ideas are then extended to allow matrices of the form (4.1) to be updated, whilst retaining the block structure.

It is convenient to regard the Simplex update as taking place in two stages. First a column is removed from $A$ and columns to the right are moved one place to the left. The factors of the resulting matrix are restored to triangular form. Then the matrix is extended by adding the new column as column $n$. We now review how the FM algorithm updates LU factors under these transformations.

When a column is removed from $A$ the resulting matrix has factors in which the corresponding column has been removed from $U$. For example if column 2 is removed from a $5 \times 5$ matrix $A$, the resulting matrix has factors

$$\begin{bmatrix} a_{11} & a_{13} & a_{14} & a_{15} \\ a_{21} & a_{23} & a_{24} & a_{25} \\ a_{31} & a_{33} & a_{34} & a_{35} \\ a_{41} & a_{43} & a_{44} & a_{45} \\ a_{51} & a_{53} & a_{54} & a_{55} \end{bmatrix} = \begin{bmatrix} 1 & & & & \\ l_{21} & 1 & & & \\ l_{31} & l_{32} & 1 & & \\ l_{41} & l_{42} & l_{43} & 1 & \\ l_{51} & l_{52} & l_{53} & l_{54} & 1 \end{bmatrix} \begin{bmatrix} u_{11} & u_{13} & u_{14} & u_{15} \\ & u_{23} & u_{24} & u_{25} \\ & u_{33} & u_{34} & u_{35} \\ & & u_{44} & u_{45} \\ & & & u_{55} \end{bmatrix}.$$

The aim of the FM method is to eliminate the resulting subdiagonal elements of $U$ in a stable way. Without loss of generality we assume that column 1 of $A$ has been removed, and $A$ and $U$ are renumbered to reflect this change. The first stage of the method determines an elementary matrix $B$ which eliminates $u_{21}$ in the equation

$$\begin{bmatrix} a_{11} \\ a_{21} \end{bmatrix} = \begin{bmatrix} 1 & \\ l_{21} & 1 \end{bmatrix} B^{-1} B \begin{bmatrix} u_{11} \\ u_{21} \end{bmatrix}$$

whilst retaining the unit triangular structure of $L$. If $u_{11}$ is not zero, the matrix

$$B = \begin{bmatrix} 1 & \\ r & 1 \end{bmatrix} \tag{5.1}$$

can be used, where $r = -u_{21}/u_{11}$. The is referred to as the "no-perm" operation. Alternatively, an interchange may first be made to rows 1 and 2 of $A$, which also interchanges rows 1 and 2 of $L$. Then $B$ may be made up of the product of two elementary matrices, as defined by the equation

$$\begin{bmatrix} a_{21} \\ a_{11} \end{bmatrix} = \begin{bmatrix} l_{21} & 1 \\ 1 & \end{bmatrix} \begin{bmatrix} & 1 \\ 1 & -l_{21} \end{bmatrix} \begin{bmatrix} 1 & \\ -s & 1 \end{bmatrix} \begin{bmatrix} 1 & \\ s & 1 \end{bmatrix} \begin{bmatrix} l_{21} & 1 \\ 1 & \end{bmatrix} \begin{bmatrix} u_{11} \\ u_{21} \end{bmatrix} \tag{5.2}$$

where $s = -u_{11}/\Delta$ and $\Delta = l_{21}u_{11} + u_{21}$. The outer pair of matrices in $B^{-1}B$ induce a unit submatrix in $L$, and the inner pair eliminate the resulting subdiagonal element of $U$. This is referred to as the "perm" operation. One of these possibilities is always valid. To minimize a bound on growth in the factors, Fletcher and Matthews essentially use the no-perm operation when the test

$$|\Delta| \le |u_{11}| \max(1, |l_{21}|) \tag{5.3}$$

is satisfied.

Once $B$ is determined, it is applied from the left to rows 1 and 2 of $U$. Also $B^{-1}$ is applied from the right to columns 1 and 2 of $L$. This requires two saxpy-like operations in the no-perm case and four in the perm case, and is the major cost of the stage. The same process is repeated in subsequent stages to eliminate the remaining subdiagonal elements of $U$. The total cost of the elimination is $O(n^2)$, with the constant depending on which column is removed and the extent to which perm operations are used. Finally a new column, $\mathbf{a}_n$ say, is added to $A$ and the column $\mathbf{u}_n$ to be added to $U$ is determined by solving the triangular system

$$L\mathbf{u}_n = \mathbf{a}_n, \tag{5.4}$$

where $L$ is the lower triangular factor resulting from the elimination stages. Solving this equation is a significant proportion of the cost of the whole update.

The means by which the FM algorithm may be adapted to update LIU factors is now considered. The same stages are performed and the same matrices $B$ are determined at each stage. The same test (5.3) is used to decide on a perm or no-perm operation. The differences are itemized in the following list.

- $B$ operations are applied from the left to $\mathbb{L}$, rather than $B^{-1}$ being applied to the right of $L$.
- The current matrix $U$ as the elimination proceeds is always available implicitly from the equation $\mathbb{L}A = U$, using the current copy of $\mathbb{L}$. Thus, although the $u_{21}$ elements are directly available from $D$, the $u_{11}$ elements must be generated by a scalar product between the relevant row of $\mathbb{L}$ and column of $A$.
- No operations on $U$ need be performed. The new diagonal elements of $D$ are just the pivot elements (either $u_{11}$ or $\Delta$) on the elimination stages.
- The element $l_{21}$ of $L$ is the negative of the corresponding element of $\mathbb{L}$.
- For the perm operation, the row permutation in $L$ corresponds to a column permutation in $\mathbb{L}$.
- The final triangular substitution (5.4) to determine $\mathbf{u}_n$ is not required, only $d_n = \mathbf{l}_n^T \mathbf{a}_n$ is needed.

Thus a no-perm operation simply applies the matrix $B$ in (5.1) from the left to rows 1 and 2 of the equation $\mathbb{L}A = U$. In the perm operation, columns 1 and 2 of $\mathbb{L}$ are first interchanged. Then an elementary operation is applied from the left, which induces a unit submatrix in $\mathbb{L}$. Finally another elementary operation is applied from the left to eliminate the resulting subdiagonal of $U$.

The cost of the algorithm in the case of LIU factors is $O(n^2)$, as for LU factors, but there are some differences. There is a considerable saving in not having to update $U$ in the LIU case, but this is partially offset by the need to calculate $u_{11}$ at each stage by a scalar product. Also more arithmetic operations are required to update $\mathbb{L}$ than $L$, except when it is column 1 that is removed from $A$. However, not having to solve (5.4) is a considerable saving for the LIU algorithm, and overall I would expect the LIU algorithm to be noticeably more efficient in practice.

Updating strategies can also be determined when $A$ is kept in the block form (4.1). If a non-unit vector is added or removed then the procedure described above for LIU factors also applies. Therfore only the cases in which a unit vector is added or removed need to be considered.

When a unit vector, $\mathbf{e}_{n-m}$ say, is removed from $A$ in (4.1), it may be the case that the vector $\mathbf{s}$ defined by solving

$$A^T\mathbf{s} = \mathbf{e}_{n-m} \tag{5.5}$$

is available. This is often the case in LP calculations. One way to extend $\mathbb{L}$ is to move the unit element in row $n-m$ of $\mathbf{s}$ to row $n$, and shift up the intermediate elements. The same permutation is made to rows of $\mathbb{L}$ and columns of $A$. Then, using the new values, column $n$ of $A$ is deleted and row $n$ of $\mathbb{L}$ is replaced by $\mathbf{s}$. It follows from the orthogonality conditions implied by (5.5) that $\mathbb{L}A$ is still upper triangular. Unfortunately it is possible that growth in the elements of $\mathbf{s}$ and hence $\mathbb{L}$ can occur with this method. A more stable alternative is simply to extend $A_1$ by moving the partition lines in (4.1) one column to the left and one row upwards. Then the procedure described earlier in this section may be used to update the LIU factors of $A_1$ when its first column is deleted. However, if $\mathbf{s}$ is available, it is worth using the more simple alternative if no significant growth is observed.

The main differences in procedure occur when a unit vector, $\mathbf{e}_p$ say, is added to $A$. To introduce this unit vector as the last column of $A$ would not conform to the structure defined in (4.1) in which the unit columns are to the left of the non-unit columns. The correct structure can be maintained by moving row $p$ of $A$ into the $A_2$ partition, and including a unit column in $A$. Hence what is required is a procedure for updating the LIU factors of a matrix $A$ when a *row* is removed from $A$. Now it is possible (Matthews [14]) to adapt the Fletcher–Matthews method so as to update LU factors when a row of $A$ is removed. What is described here is a modified form of this procedure which allows the LIU factors to be updated. It is assumed that a column has previously been deleted from $A$, leaving $A$ and $U$ as $n \times (n-1)$ matrices, and $\mathbb{L}$ as an $n \times n$ matrix, with $\mathbb{L}A = U$. The row to be deleted from $A$ is denoted by $p$. The procedure takes place in a number of stages, indexed by $p, p+1, \ldots, n-1$. It is convenient to regard each stage as starting off by interchanging rows $p$ and $p+1$ of $A$ and columns $p$ and $p+1$ of $\mathbb{L}$. The net effect over all stages is to cyclically permute row $p$ of $A$ to row $n$, when it can then be deleted.

The generic procedure within a stage is now described, in the case that $p = 1$. After the initial interchange, the leading submatrix satisfies the equation

$$\begin{bmatrix} & l_{11} \\ 1 & l_{21} \end{bmatrix} \begin{bmatrix} a_{21} & a_{22} \\ a_{11} & a_{12} \end{bmatrix} = \begin{bmatrix} u_{11} & u_{12} \\ & u_{22} \end{bmatrix}. \tag{5.6}$$

In fact $l_{11} = 1$ holds at the first stage, but this is not necessarily the case at subsequent stages, although $l_{11} \neq 0$ since $\mathbb{L}$ is always nonsingular. Having $l_{11} \neq 1$ enables an additional scaling operation to be avoided. The elements $u_{11}$ and $u_{22}$ in (5.6) are available as $d_1$ and $d_2$ respectively, but $u_{12}$ must be evaluated at the start of the stage by a scalar product between the relevant row of $\mathbb{L}$ and column of $A$. It is required to multiply equation (5.6) on the left by a matrix $B$ in order to restore a lower triangle in $\mathbb{L}$ and retain the upper triangle in $U$. A straightforward possibility is to choose $B$ as the elementary matrix

$$B = \begin{bmatrix} r & 1 \\ & 1 \end{bmatrix} \tag{5.7}$$

where $r = -l_{21}/l_{11}$, and this is referred to as the "no-perm" operation. This fails when $l_{21} = 0$ as the resulting lower triangular matrix has zero on the diagonal. The alternative choice defines $B$ as the product of two elementary matrices

$$B = \begin{bmatrix} 1 & \\ s & 1 \end{bmatrix} \begin{bmatrix} r & 1 \\ 1 & \end{bmatrix} \tag{5.8}$$

where $s = -u_{12}/\Delta$ and $\Delta = u_{22} - l_{21}u_{12}/l_{11}$, and this is referred to as the "perm" operation. In this case a subsequent column interchange in $A$ and $U$ is used to make the leading submatrix in $U$ upper triangular. This operation fails if $\Delta = 0$. A test analogous to (5.3) is to choose the no-perm operation when

$$|\Delta| \leq \left|\frac{l_{21}}{l_{11}}\right| \max(|u_{11}|, |u_{12}|). \tag{5.9}$$

The same procedure is also applied at subsequent stages. At stages other than $k = p = 1$ the operation defined by $B$ is also applied to the other elements in rows $p$ and $p+1$ of $\mathbb{L}$. This is the dominant cost of the updating process, which therefore requires $O(n^2)$ arithmetic operations in all.

# 6 Other implicit factorizations

An interesting observation is that the algorithms (2.15) and (2.16) for respectively solving $A\mathbf{x} = \mathbf{b}$ and $A^T\mathbf{x} = \mathbf{b}$ do *not* require $\mathbb{L}$ to be triangular, only $U$. Hence another practicable implicit factorization would be to have factors

$$SA = U \tag{6.1}$$

where $S$ is square. These are referred to as SIU factors. As before, the diagonal elements of $U$ are available as a matrix $D$, but the rest of $U$ is implicit. An advantage of this format is that Simplex updates take a more simple form in which only one elementary operation per stage is required. This is described by reference to the case in which column 2 has been deleted from a $4 \times 4$ matrix $A$, leaving a matrix $A'$. The first stage of the updating process is to eliminate $u_{33}$ from the SIU factors. If $|u_{23}| \geq |u_{33}|$ then the operation

$$\begin{bmatrix} 1 & & & \\ & 1 & 0 & \\ & r & 1 & \\ & & & 1 \end{bmatrix} \begin{bmatrix} s_{11} & s_{12} & s_{13} & s_{14} \\ s_{21} & s_{22} & s_{23} & s_{24} \\ s_{31} & s_{32} & s_{33} & s_{34} \\ s_{41} & s_{42} & s_{43} & s_{44} \end{bmatrix} A' = \begin{bmatrix} 1 & & & \\ & 1 & 0 & \\ & r & 1 & \\ & & & 1 \end{bmatrix} \begin{bmatrix} u_{11} & u_{13} & u_{14} \\ & u_{23} & u_{24} \\ & u_{33} & u_{34} \\ & & u_{44} \end{bmatrix}.$$

can be used, where $r = -u_{33}/u_{23}$. Otherwise the operation

$$\begin{bmatrix} 1 & & & \\ & 0 & 1 & \\ & 1 & s & \\ & & & 1 \end{bmatrix} \begin{bmatrix} s_{11} & s_{12} & s_{13} & s_{14} \\ s_{21} & s_{22} & s_{23} & s_{24} \\ s_{31} & s_{32} & s_{33} & s_{34} \\ s_{41} & s_{42} & s_{43} & s_{44} \end{bmatrix} A' = \begin{bmatrix} 1 & & & \\ & 0 & 1 & \\ & 1 & s & \\ & & & 1 \end{bmatrix} \begin{bmatrix} u_{11} & u_{13} & u_{14} \\ & u_{23} & u_{24} \\ & u_{33} & u_{34} \\ & & u_{44} \end{bmatrix}.$$

is used, where $s = -u_{23}/u_{33}$, and the situation is somewhat reminiscent of the Bartels–Golub method [2]. Subsequent stages are used to eliminate all the subdiagonal elements of $U$. As in the previous section, the elements $u_{kk}$ are available as the elements $d_k$, but $u_{k-1,k}$ must be evaluated when needed by a scalar product between row $k-1$ of the current $S$ matrix and column $k$ of $A$. Once $U$ has been returned to triangular form, a new column $\mathbf{a}_n$ can be added to $A$. As $U$ is implicit, all that needs to be done is to calculate $d_n = \mathbf{s}_n^T \mathbf{a}_n$, where $\mathbf{s}_n^T$ denotes row $n$ of $S$.

A related possibility is to update implicit factors

$$QA = U \tag{6.2}$$

(QIU factors) in which $Q$ is an orthogonal matrix. In this case the elimination of a subdiagonal element of $U$ can be done using Givens' rotations. This is illustrated by

$$\begin{bmatrix} 1 & & & \\ & c & s & \\ & s & -c & \\ & & & 1 \end{bmatrix} \begin{bmatrix} q_{11} & q_{12} & q_{13} & q_{14} \\ q_{21} & q_{22} & q_{23} & q_{24} \\ q_{31} & q_{32} & q_{33} & q_{34} \\ q_{41} & q_{42} & q_{43} & q_{44} \end{bmatrix} A' = \begin{bmatrix} 1 & & & \\ & c & s & \\ & s & -c & \\ & & & 1 \end{bmatrix} \begin{bmatrix} u_{11} & u_{13} & u_{14} \\ & u_{23} & u_{24} \\ & u_{33} & u_{34} \\ & & u_{44} \end{bmatrix},$$

where $c = u_{23}/\sqrt{u_{23}^2 + u_{33}^2}$ and $s = u_{33}/\sqrt{u_{23}^2 + u_{33}^2}$.

Both these methods require $n^2 + O(n)$ storage, so both this and the operations count are greater than for LIU factors. However, the numerical stability of these methods is likely to be superior. It can be assumed that numerical stability is correlated with any growth that occurs in the matrices $\mathbb{L}$, $S$ or $Q$. Of course the elements of $Q$ are bounded in modulus by 1, so there is no growth and this can be expected to give the most stable method. The growth in $S$ is limited by a factor of 2 per stage, which is a better bound than can be obtained for the Fletcher–Matthews method, and hence presumably for the methods derived in Section 5. Thus it would be expected that the SIU method is more stable than the LIU method. In practice it is likely that the actual growth is much less than is predicted by the bound. It is also observed that the determinants of $\mathbb{L}$, $S$ and $Q$ are all $\pm 1$. Some experiments which attempt to quantify the cost and stability of these methods are described in the next section.

Another implicit factorization, which is quite different from the methods discussed in this paper, is to have QR factors

$$A = QR \tag{6.3}$$

in which $A$ and $R$ are stored, and $Q$ is defined implicitly by $Q = AR^{-1}$. This only requires $\sim \frac{1}{2}n^2$ storage and so is comparable with LIU factors in this respect. It is also possible to update the factors using Givens' rotations when columns are added to or removed from $A$ (Gill and Murray [10], Siegel [17]). A disadvantage is that two solves with $R$ are needed to solve $A\mathbf{x} = \mathbf{b}$. Thus the forward error in $\mathbf{x}$ behaves like $\kappa(A)^2\varepsilon$ rather than $\kappa(A)\varepsilon$ for LIU factors, so the method is more seriously affected by ill-conditioning. For this reason this method is little used, although there is some evidence (Björck [5]) that iterative refinement can improve matters.

# 7 Numerical results and Conclusions

To quantify the relative costs and numerical stability of the different implicit factorizations, a range of calculations on eleven LP problems from the SOL test set with $n$ up to 688 has been carried out. These calculations involve a mixture of Simplex updates to $A$ and solves with $A$ and $A^T$. These calculations dominate the cost of solving the LP problems and so provide an overall measure of the efficiency of the different implicit factorization schemes. The calculations have been repeated with different types of prescaling strategies and the use or non-use of steepest edge pricing. The coefficient matrices in the LP problems, and hence the columns of $A$, are sparse and are stored in compact form.

Not much is gained by reporting the detailed outcome of all these tests (tables are available from me on request) but the following statistics provide an overview.

**Computer Time**

$$\frac{\text{LIU method}}{\text{SIU method}} \qquad \text{typical range} \qquad 0.7 \sim 0.8$$

$$\frac{\text{QIU method}}{\text{SIU method}} \qquad \text{typical range} \qquad 1.5 \sim 2.0$$

The first of these comparisons reflects the improvement in efficiency obtained by working with a triangular matrix $\mathbb{L}$ as against a square matrix $S$, moderated by the extra operations required by a perm elimination and the added complexity of manipulating triangular matrices. The overall effect is about what might have been predicted. The second comparison indicates the additional cost of using Givens' rotations rather than elementary elimination operations, and the ratio of cost is somewhat larger than I would have expected.

Two different indicators of numerical stability are given. The first is to monitor the maximum growth in $\mathbb{L}$ and $S$ over *all* LP iterations.

**Maximum growth**

| | | |
|---|---|---|
| LIU method | typical range | $100 \sim 2000$ |
| SIU method | typical range | $5 \sim 50$ |

Of course no growth in $Q$ is obtained. It should be stressed that these figures represent the worst that happens during the LP calculation and that very large growth only arises on the occasional iteration, so that the average growth is much less. Scaling and the use of steepest edge pricing also tend to reduce the maximum growth that occurs. Nonetheless one would expect that when large growth does occur, then a corresponding growth in backward error would be obtained which would persist through subsequent updates.

Another indication of numerical stability is to calculate the accumulated error, by evaluating the residual of the equation $\mathbb{L}A = U$ (or $SA = U$ or $QA = U$) at the end of each LP calculation. The outcome is very uneven but an overall impression has been obtained by averaging (in log space) the maximum residual for each LP problem. Scaling columns of $A$ to have unit length has a significant effect, so results for unscaled problems are given separately. The results are obtained on a SUN ELC in double precision, for which the machine precision is $\varepsilon \simeq 1.1_{10} - 16$.

**Accumulated error**

| | LIU | SIU | QIU |
|---|---|---|---|
| Scaled problems | $1.8_{10} - 14$ | $1.5_{10} - 14$ | $4.3_{10} - 15$ |
| Unscaled problems | $2.4_{10} - 12$ | $1.6_{10} - 12$ | $2.5_{10} - 13$ |

In fact the SIU method did somewhat better than these results indicate, as the averages include 3 very poor results (out of 44) for the scaled problems. If these results are removed from the comparison then the average improves to $5.2_{10} - 15$ which is almost as good as the QIU method. Also, if a bad result for the SIU method is deleted from the 22 runs on scaled problems, the average improves to $8.4_{10} - 13$. Given that the variance of the results is large, the most that can be concluded is that the LIU method is about a factor of 10 worse on average than the QIU method, with the SIU method somewhere in the middle.

Given the much higher maximum growth for the LIU method, it is somewhat surprising that a more substantial discrepancy in the accumulated error is not obtained. The reason may be that when large growth occurs, the associated large errors are confined to the column that has been introduced. When this column is subsequently removed, it may be that the associated large error is also removed. Similar conclusions were reached in [7] in regard to updates of LU factors.

Overall the LIU method requires only half the storage of the SIU and QIU methods, and it also performs best in regard to computing time. On average it is about a factor of 10 worse in regard to accuracy than the QIU method. Since the QIU method obtains accuracy close to the machine precision, this seems to be a small price to pay for the better performance. Similar conclusions could be expected if the method were compared with regular (dense) LU or QR factors.

# References

[1] Abaffy, J., C. Broyden and E. Spedicato (1984) "A class of direct methods for linear systems", *Numer. Math.* **45**, 361–376.

[2] Bartels, R.H. (1971) "A stabilization of the simplex method", *Numer. Math.* **16**, 414–434.

[3] Benzi, M. and C.D. Meyer (1994) "An explicit preconditioner for the conjugate gradient method", *Proceedings of the Lanczos Centenary Conference* (J. Brown, M. Chu, D. Ellison and R. Plemmons, eds), SIAM Publications (Philadelphia).

[4] Benzi, M. and C.D. Meyer (1995) "A direct projection method for sparse linear systems", *SIAM J. Sci. Comput.* **16**, 1159–1176.

[5] Björck, Å. (1987) "Stability analysis of the method of the method of semi-normal equations for least squares problems", *Linear Algebra Applns* **88/89**, 31–48.

[6] Enderson, D. and A. Wassyng (1978) "A new method for the solution of $Ax = b$", *Numer. Math.* **29**, 287–289.

[7] Faddeev, D.K. and V.N. Faddeeva (1963) *Computational Methods of Linear Algebra*, W.H. Freeman, San Francisco.

[8] Fletcher, R. and S.P.J. Matthews (1984) "Stable modification of explicit LU factors for simplex updates", *Math. Progr.* **30**, 367–284.

[9] Fox, L., H.D. Huskey and J.H. Wilkinson (1948) "Notes on the solution of algebraic linear simultaneous equations", *Quart. J. Mech. Appl. Math.* **1**, 149–173.

[10] Gill, P.E. and W. Murray (1973) "A numerically stable form of the simplex algorithm", *Linear Algebra Applns* **7**, 99–138.

[11] Hegedüs, C.J. (1986) "Newton's recursive interpolation in $\mathbb{R}^n$", *Colloquia Math. Soc. János Bolyai* **50** 605–623.

[12] Hestenes, M.R. and E. Stiefel (1952) "Methods of conjugate gradients for solving linear systems", *J. Res. NBS* **49**, 409–436.

[13] Householder, A.S. (1955) "Terminating and non-terminating iterations for solving linear systems", *J. SIAM* **3**, 67–72.

[14] Matthews, S.P.J. (1984) "Matrix algebra and other aspects of linear and quadratic programming", Ph.D. Thesis, Dept. of Math. Sci., Univ. of Dundee, Scotland.

[15] Pietrzykowski, T. (1960) "Projection method", *Zak. Apar. Mat. Polskiej Acad. Nauk*, Praca A8.

[16] Purcell, E.W. (1953) "The vector method of solving simultaneous linear equations", *J. Math. Phys.* **32**, 180–183.

[17] Siegel, D. (1992) "Implementing and modifying Broyden class updates for large scale optimization", Report DAMTP 1992/NA12, Univ. of Cambridge.

[18] Sloboda, F. (1978) "A parallel projection method for linear algebraic systems", *Apl. Mat. Ceskosl. Akad. Ved.* **23**, 185–198.

[19] Stewart, G.W. (1973) "Conjugate direction methods for solving linear systems", *Numer. Math.* **21**, 283–297.

[20] Stewart, G.W. (1973) *Introduction to Matrix Computations*, Academic Press (New York).

[21] Suhl, U.H. (1994) "MOPS – Mathematical Optimization System", *European J. Operational Research* **72**, 312–322.

[22] Tuma, M. (1993) "Solving sparse unsymmetric sets of linear equations based on implicit Gauss projection", Tech. Report 555, Inst. of Comp. Sci., Czech Acad. Sci., Prague.

[23] Wassyng, A. (1982) "Solving $Ax = b$: a method with reduced storage requirements", *SIAM J. Num. Anal.* **19**, 197–204.

[24] Wilkinson, J.H. (1965), *The Algebraic Eigenvalue Problem*, Oxford University Press (Oxford).

# Optimization Environments and the NEOS Server

*William Gropp*
*Jorge J. Moré*
Mathematics and Computer Science Division, Argonne National Laboratory, Argonne, IL 60439-4844, USA[1]

## 1 Introduction

In an ideal computational environment the user would formulate the optimization problem and obtain results without worrying about computational resources. Unfortunately this ideal environment is not possible because if sufficient care is not given to the formulation, a reasonable problem may become untractable. Even with an appropriate formulation, obtaining the solution of difficult optimization problems requires sophisticated optimization software and access to large-scale computational resources. Modeling three-dimensional physical processes by systems of differential equations gives rise to optimization problems that require access to substantial computational resources. Discrete and global optimization problems are also in this category.

We are interested in the development of problem-solving environments that simplify the formulation of optimization problems, and the access to computational resources. Once the problem has been formulated, the first step in solving an optimization problem in a typical computational environment is to identify and obtain the appropriate piece of optimization software. The software may be available from a mathematical software library, or may need to be bought and installed. In some cases the software is in the public domain, and available from a site on the Internet. Once the software has been installed and tested in the local environment, the user must read the documentation and write code to define the optimization problem in the manner required by the software. Typically, Fortran or C code must be written to define the problem, compute function values and derivatives, and specify sparsity patterns. Finally, the user must debug, compile, link, and execute the code.

[1]This work was supported by the Mathematical, Information, and Computational Sciences Division subprogram of the Office of Computational and Technology Research, U.S. Department of Energy, under Contract W-31-109-Eng-38, and by a grant of Northwestern University to the Optimization Technology Center.

The Network-Enabled Optimization System (NEOS) is an Internet-based service providing information, software, and problem-solving services for optimization. The main components of NEOS are the NEOS Guide and the NEOS Server. Additional information on the various services provided by NEOS can be obtained from the home page of the Optimization Technology Center

`http://www.mcs.anl.gov/home/otc/`

The NEOS Server [11] is a novel environment that allows users to solve optimization problems over the Internet while requiring only that the user provide a specification of the problem. There is no need to download an optimization solver, write code to link the optimization solver with the optimization problem, or compute derivatives. Moreover, the NEOS Server provides an interface that is problem oriented and independent of the computing resources offered by NEOS. As long as there is an efficient way to describe the problem, the NEOS Server can provide access to a wide variety of computational services, from small clusters of workstations to any number of participating supercomputer centers.

The current version of the NEOS Server is described in Section 2. We emphasize nonlinear optimization problems, but NEOS does handle linear and nonlinearly constrained optimization problems, and solvers for optimization problems subject to integer variables are being added.

Performance issues are examined in Section 3. In particular, we provide evidence that the NEOS Server is able to solve large nonlinear optimization problems in time comparable to software with hand-coded gradients. We do not discuss the design and implementation of the Server because these issues are covered by Czyzyk, Mesnier and Moré [11].

In Section 4 we begin to explore possible extensions to the NEOS Server by discussing the addition of solvers for global optimization problems. Section 5 discusses how a remote procedure call (RPC) interface to NEOS addresses some of the limitations of NEOS in the areas of security and usability. The detailed implementation of such an interface raises a number of questions, such as exactly how the RPC is implemented, what security or authentication approaches are used, and what techniques are used to improve the efficiency of the communication. These questions are not discussed here. Instead, we outline some of the issues in network computing that arise from the emerging style of computing used by NEOS.

## 2 The NEOS Server

The NEOS Server provides Internet access to a library of optimization solvers with user interfaces that shield the user from the optimization software. The user needs only to describe the optimization problem; all additional information required by the optimization solver is determined automatically.

The NEOS approach offers considerable advantages over a conventional environment for solving optimization problems. Consider, for example, how NEOS solves an optimization problem of the form

$$\min \{ f(x) : x \in \mathbb{R}^n \},$$

where $f : \mathbb{R}^n \to \mathbb{R}$ is partially separable, that is, $f$ can be written as

$$f(x) = \sum_{i=1}^{n_f} f_i(x),$$

where each element function $f_i$ only depends on a few components of $x$, and $n_f$ is the number of element functions. Algorithms and software that take advantage of partial separability have been developed for various problems (for example, [23, 24, 25, 26, 9]), but this software requires that the user provide the gradient of $f$ and the partial separability structure (a list of the dependent variables for each element function $f_i$).

The NEOS solvers for partially separable problems require that the user specify the number of variables `n`, a subroutine `initpt(n,x)` that defines the starting point, and a subroutine `fcn(n,x,nf,fvec)` that evaluates the element functions. Since there is no need to provide the gradient or the partial separability structure, the user can concentrate on the specification of the problem. Changes to the `fcn` subroutine can be made and tested immediately; the advantages in ease of use are considerable.

The NEOS solvers for the bound constrained problem

$$\min \{ f(x) : x_l \le x \le x_u \}$$

and the nonlinearly constrained optimization problem

$$\min \{ f(x) : x_l \le x \le x_u, \ c_l \le c(x) \le c_u \}$$

also make use of partial separability. The bound constrained problem is specified by a subroutine that specifies the bounds $x_l$ and $x_u$, while for the nonlinearly constrained problem we also need a subroutine that specifies the constraint bounds $c_l$ and $c_u$, and the nonlinear function $c : \mathbb{R}^n \mapsto \mathbb{R}^m$. Specifying this information is done by additional subroutines. The bounds $x_l$ and $x_u$ are specified by the subroutine `xbound(n,xl,xu)`, the constraint bounds $c_l$ and $c_u$ are specified by the subroutine `cbound(m,cl,cu)`, and the nonlinear function $c : \mathbb{R}^n \mapsto \mathbb{R}^m$ is specified by `cfcn(m,x,c)`.

We have mentioned nonlinear optimization solvers, but NEOS contains solvers in other areas. At present we have solvers in the following areas:

Unconstrained optimization
Bound constrained optimization
Nonlinearly constrained optimization
Complementarity problems
Linear network optimization
Linear programming
Stochastic linear programming

The addition of solvers in other areas is not difficult; indeed, NEOS was designed so that solvers in a wide variety of optimization areas could be added easily.

We provide Internet users with the choice of three interfaces for submitting problems: e-mail, the NEOS Submission tool, and the NEOS Server Web interface. The interfaces are designed so that problem submission is intuitive and requires the minimal amount of information. The interfaces differ only in the way that information is specified and passed to the NEOS Server.

The e-mail interface is relatively primitive, but useful because most users have easy access to e-mail. Information on the available services and on the format used to submit problems via e-mail can be obtained by sending the mail message help to

`neos@mcs.anl.gov`

Users interested in the Web interface should visit the URL

**`http://www.mcs.anl.gov/home/otc/Server/`**

This URL has links to all the solvers in the library, as well as pointers to other NEOS information, in particular the NEOS Guide. In the remainder of this section we examine the NEOS Submission tool.

The NEOS Submission tool provides a high-speed link to the NEOS Server via sockets. Once this tool is installed, the user has access to all the services provided by the NEOS Server. Users may download the Submission tool from the URL

**`http://www.mcs.anl.gov/otc/Server/submission_tool.html/`**

Installation of the Submission tool is immediate provided that Perl [28] has been installed properly. If the installation fails, the usual remedy is to run the Perl **h2ph** script that changes C header files into Perl header files. Running the **h2ph** script is simple but must be done by the installer of Perl, who is usually the system administrator.

Submission of problems via the NEOS Submission tool is simple. The user must first choose the type of optimization problem. Once an area is selected, the user must choose a solver. This selection process is done via drop-down menus typical of well-designed user interfaces.

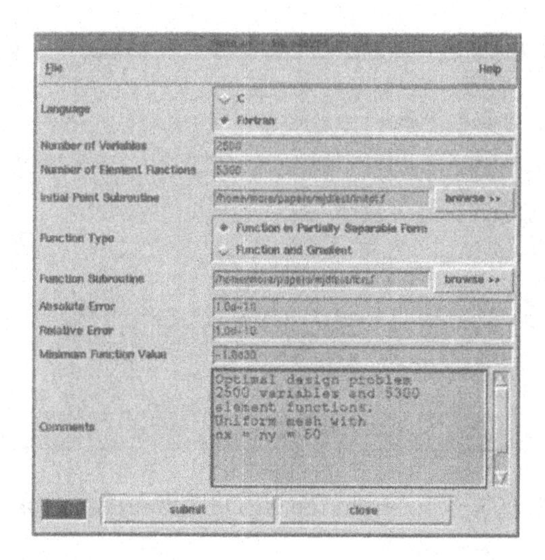


**Figure 2.1** The NEOS submission form for vmlm.

The optimization problem is specified via a submission form. For example, Figure 2 shows the NEOS Submission form for the vmlm solver of unconstrained optimization problems. Solvers in each area have a submission form that is appropriate for that area.

For the vmlm solver the user needs to specify the language used to submit the problem (Fortran or C), the number of variables $n$, the number of partially separable functions $n_f$, and the files for the initial point and function evaluation subroutines. Browse buttons are available to ease the specification of the various files. An advantage of this interface is that, unlike the Web interface, the subroutines can be in files that reside in the user's local file space.

The general philosophy of the NEOS solvers is that problem submission should be intuitive and require only essential information. Parameters that affect the progress of the algorithm are not required but can be specified, for example, by a specification file. The vmlm solver allows the user a choice of tolerances, but for most problems the defaults provided are adequate. The form also has room for comments, which can be used to identify the problem submission.

Once the problem is specified, the problem is submitted via the submission button at the bottom of the form (see Figure 2.1). A variety of computers, even a massively parallel processor, could be used to solve the problem; the only restriction is that the computer must run UNIX with support for TCP/IP. At present these computers are workstations that reside at Argonne National Laboratory, Northwestern University, and the University of Wisconsin.

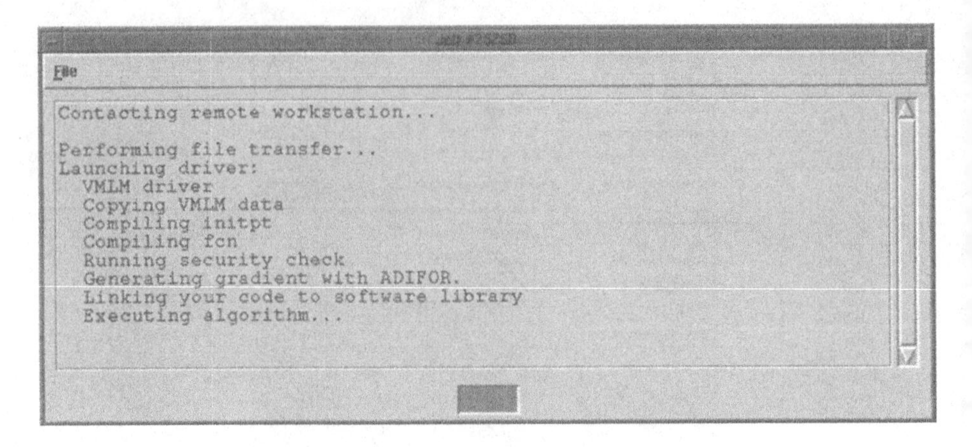


**Figure 2.2** Output from the NEOS Submission tool.

For a typical submission, the user receives information on the progress of the submission, and the solution. Figure 2.2 shows part of the output received when the problem in Figure 2.1 is submitted to NEOS. This output shows that NEOS contacts an available workstation and transfers all of the user's data to the workstation. The solver (in this case vmlm) checks the data and compiles the user's code. If any errors are found at this stage, the compiler error messages are returned to the user, and execution terminates.

If the user's code compiles correctly, automatic differentiation tools (ADIFOR [4, 3] for Fortran code) are used to generate the gradient. Once the gradient is obtained, the user's code is linked with the software library, and execution begins. Results are returned in the window generated by the NEOS Submission tool.

Interesting issues arise during the processing of the job submission that are pertinent to the development of optimization software and problem-solving environments. For example, high-quality software should check the input data, but in this case the data is the Fortran programs initpt and fcn. In general, it is not possible to check that this data is correct. At present we check only that the user function does not create any system exceptions during the evaluation of the function at the starting point. Although simple, this test catches many errors on the part of the user.

Submitting a problem to the NEOS Server does not guarantee success, but NEOS users are able to solve difficult optimization problems without worrying about many of the details that are typical in a computing environment. Even if the user has suitable optimization software, the user would need to read the documentation, write code to interface the problem with the optimization software, and then debug this code. The user would also have to write and debug code for the gradient—a nontrivial undertaking in most cases.

# 3 Performance

The NEOS solvers for partially separable problems are able to solve large-scale nonlinear optimization problems while requiring only that the user provide code for the function evaluation. This ability was considered unrealistic until recently. The major obstacle was the computation of the gradient. For small-scale problems we can approximate the gradient by differences of function values, for example,

$$[\nabla f(x)]_i \approx \frac{f(x + h_i e_i) - f(x)}{h_i}, \qquad 1 \le i \le n,$$

where $h_i$ is the difference parameter and $e_i$ is the $i$th unit vector, but this approximation is prohibitive for large-scale problems because it requires $n$ function evaluations for each gradient. Approximating a gradient by differences is not only expensive but also increases the unreliability of the optimization code, since a poor choice for $h_i$ may cause premature termination of the optimization algorithm far away from the solution.

The NEOS solvers for nonlinear optimization problems use automatic differentiation tools to compute the gradients, Jacobians, and sparsity patterns required by the solvers. At present, we rely on ADIFOR [4, 3] to process Fortran code and on ADOL-C [15] to process C code.

We demonstrate the ability of NEOS to solve large-scale nonlinear optimization problems with an optimal design problem formulated by Goodman, Kohn and Reyna [14]. This optimal design problem requires determining the placement of two elastic materials in the cross section of a rod with maximal torsional rigidity. The mathematical formulation is to minimize a functional of the form

$$f_\lambda(v) = \int_{\mathcal{D}} \left\{ \psi_\lambda \left( \|\nabla v(x)\| \right) + v(x) \right\} dx,$$

over a domain $\mathcal{D}$ in $\mathbb{R}^2$, where $\psi_\lambda : \mathbb{R} \mapsto \mathbb{R}$ is a piecewise quadratic. The formulation of the optimal design problem with finite elements leads naturally to a partially separable optimization problem in $n = n_x n_y$ variables, where $n_x$ and $n_y$ are the number of interior grid points in the coordinate directions, respectively. We use the formulation in the MINPACK-2 test problem collection [1]. Additional details on the problem formulation are not important to our discussion. We need to know only that in our numerical results we consider the problem of minimizing $f_\lambda$ for a fixed value of $\lambda$; in this case $\lambda = 0.008$.

From a computational viewpoint, the most interesting feature of the code to evaluate $f_\lambda$ is that the number of floating-point operations required to evaluate $f_\lambda$ grows linearly with $n$. Ideally, we would like to solve the problem in time proportional to $n$.

We solve the optimal design problem by developing code to evaluate $f_\lambda$. In our formulation the vector $\mathbf{x}$ contains the values of the piecewise linear finite element approximation, and the subroutine

```
dodc(nx,ny,x,nf,fvec,lambda)
```

evaluates the components of the partially separable function $f_\lambda$ as a function of the number of grid points and $\lambda$. The components of the partially separable function are stored in the array `fvec` of length `nf`. In this formulation `nf` is the number of elements in the finite element triangulation.

Ths subroutine `dodc` does not have the desired form for submission to NEOS, but it is quite easy to write a wrapper. For example, the results in this section were obtained with a subroutine of the form

```
fcn(n,x,nf,fvec)
```

that sets `nx` and `ny` to $n^{1/2}$ and $\lambda$ to 0.008. With this formulation we can quickly submit a series of problems to NEOS for various values of $n$.

Submission of the optimal design problem with the NEOS Submission tool is quite easy. Figure 2.1 shows the form that was used to submit the optimal design problem. In Figure 2.1 we were using $n = 2500$, but the form can be used for other values of $n$ by changing the number of variables and the number of element functions.

Table 3.1 shows the timings (in seconds) and the number of function evaluations needed to solve an optimal design problem with the vmlm solver. We provide information for the case when the user only provides the function in partially separable form and for the case when the user provides the function and gradient. These results were obtained on a Sparc 10 with 96MB of memory.

| | Function | | | Function and Gradient | | |
|---|---|---|---|---|---|---|
| $n$ | `niters` | `nfgev` | `time` | `niters` | `nfgev` | `time` |
| 2,500 | 230 | 237 | 139 | 232 | 239 | 22 |
| 10,000 | 427 | 436 | 1042 | 427 | 433 | 165 |
| 40,000 | 865 | 871 | 8399 | 877 | 885 | 1461 |

**Table 3.1** Performance of the NEOS solver vmlm.

There are two important points to notice in the results in Table 3.1. The main point is that these results show that the time per function evaluation increases linearly with $n$. This is to be expected for this problem when the user provides both the function and the gradient, but it is remarkable that this also holds for the case when the user only provides the function. The techniques [2] used to achieve these results make essential use of the partial separability of the function.

Another important point about the results in Table 3.1 is that there is a penalty factor of six in the timings when only providing the function. If we had used a standard difference approximation to the gradient, there would have been a performance penalty of about $n$, which is prohibitive for these problems. We also note that for these results, vmlm used ADIFOR with the sparse option. This strategy is far from optimal; with the hybrid strategy of Bouaricha and Moré [7] the performance penalty is reduced to a factor of two.

Finally, we note that the number of function evaluations needed to solve the problem grows as a function of $n^{1/2}$. However, this is all that can be expected from a limited-memory variable metric method.

The main point that should be drawn from the results in Table 3.1 is that the NEOS Server combines an intuitive user interface, automatic differentiation tools, and optimization algorithms into a powerful problem-solving tool. We want to improve the NEOS Server by extending the range of problems that can be solved, but we also want to improve the interface. These issues will be examined in the next two sections.

# 4 Global optimization problems

We want to extend the NEOS Server capabilities to global optimization problems. In general, these problems are attacked by algorithms that require a large number of loosely coupled processors. We are interested, in particular, in problems that arise in connection with the determination of protein structures. As a specific instance of the type of problem under consideration, we consider distance geometry problems.

Distance geometry problems for the determination of protein structures are specified by a subset $\mathcal{S}$ of all atom pairs and by the distances between atoms $i$ and $j$ for $(i,j) \in \mathcal{S}$. In practice, lower and upper bounds on the distances are given instead of precise values. The distance geometry problem with lower and upper bounds is to find a set of positions $x_1, \ldots, x_m$ in $\mathbb{R}^3$ such that

$$l_{i,j} \le \|x_i - x_j\| \le u_{i,j}, \qquad (i,j) \in \mathcal{S}, \tag{4.1}$$

where $l_{i,j}$ and $u_{i,j}$ are lower and upper bounds on the distances, respectively. Recent reviews of the application of distance geometry problems to protein structure determination can be found in Havel [17, 18], Torda and van Gunsteren [27], Kuntz, Thomason and Oshiro [19], Brünger and Nilges [8], and Blaney and Dixon [5].

A standard formulation of the distance geometry problem (4.1), suggested by Crippen and Havel [10], is to find the global minimum of the function

$$f(x) = \sum_{i,j \in \mathcal{S}} p_{i,j}(x_i - x_j),$$

where the pairwise function $p_{i,j} : \mathbb{R}^n \mapsto \mathbb{R}$ is defined by

$$p_{i,j}(x) = \min{}^2 \left\{ \frac{\|x\|^2 - l_{i,j}^2}{l_{i,j}^2}, 0 \right\} + \max{}^2 \left\{ \frac{\|x\|^2 - u_{i,j}^2}{u_{i,j}^2}, 0 \right\}.$$

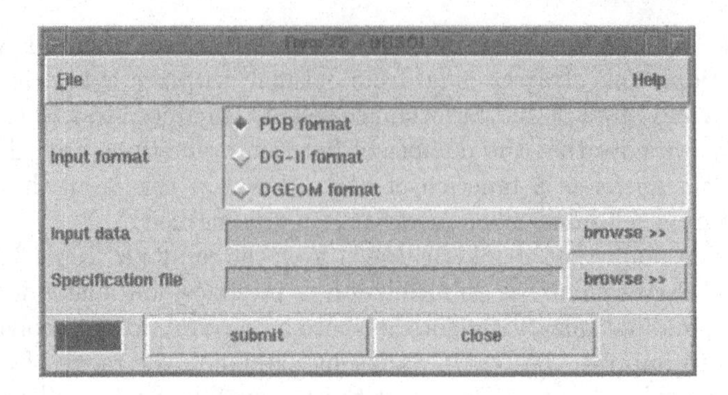


**Figure 4.1** The NEOS submission form for dgsol.

Clearly, $x = \{x_1, \ldots, x_m\}$ solves the distance geometry problem if and only if $x$ is a global minimizer of $f$ and $f(x) = 0$.

Distance geometry problems can be specified by a small amount of data. A reasonable specification is to provide the number of atoms $m$, the size of the set $\mathcal{S}$, and the vectors

$$(i, j, l_{i,j}, u_{i,j}), \qquad (i, j) \in \mathcal{S}. \tag{4.2}$$

In a practical setting this data would be accompanied by additional information (for example, the type of atoms), but for our purposes we would need only (4.2).

Finding a global minimizer of $f$ for the distance geometry problem is NP-hard even in the special case where $l_{i,j} = u_{i,j}$. Finding approximate minimizers is also NP-hard; that is, the decision problem of finding $x \in \mathbb{R}^n$ such that $f(x) \leq \epsilon$ for given $\epsilon > 0$ is NP-hard. At a more practical level, we note that the function $f$ has a large number of distinct minima, which seems to grow exponentially with the number of atoms $m$. Moreover, in our experience, even problems with a small number of atoms can be quite difficult. See Moré and Wu [21, 22] for additional information.

Several general approaches to the solution of global optimization problems could be used to solve distance geometry problems: multistart techniques, simulated annealing, and genetic algorithms. There are also specific techniques for distance geometry problems, in particular, the embed algorithm (see Crippen and Havel [10], and Havel [17, 18]), and the dgsol algorithm of Moré and Wu [22].

We have designed the NEOS Server so that additional solvers can be installed easily, with little intervention by the Server administrators. Given a solver, the user is given an interface that is intuitive and simple. For distance geometry problems we could use, for example, the submission form in Figure 4.1 that requires the distance geometry data (4.2).

Registering a new solver with the NEOS Server automatically updates the library. Submission forms, in particular, are created from configuration files specified by the software administrator. Each line of the configuration file specifies an entry in the submission form. The form in Figure 4.1, for example, requires only a three-line configuration file. For additional details on the registration process, see Czyzyk, Mesnier and Moré [11].

The main difficulty that must be faced in extending the NEOS Server to global optimization problems is the scheduling of jobs. With the current design, jobs are allocated sequentially to the first available machine, but we could use Condor [20], a distributed resource management system developed at the University of Wisconsin, to schedule jobs on heterogeneous clusters of workstations. NEOS is already using Condor to solve nonlinear complementarity problems (see Ferris, Mesnier and Moré [12]), so the transition to global optimization problems is feasible. Other research projects that are addressing the use of large-scale computing resources include Globus [13] and Legion [16].

The approach to solving global optimization problems with NEOS offers considerable advantages. Solving global optimization problems in a large-scale computing center would force the user to learn scheduling policies, job submission methods, and installation-specific software libraries. To some extent, such requirements are unavoidable; each center represents a significant investment and often must serve a particular user community. Even for users who must use these centers, the significant effort involved in learning the idiosyncrasies of each facility often limit the ability to switch between centers. In contrast, the NEOS Submission tool provides an interface that is problem oriented and independent of the computing resources that are offered by the NEOS Server.

# 5 Distributed computing

The NEOS Server is first and foremost a network resource for optimization services. The current system provides convenient access to the facilities of NEOS, and, as described in Section 4, additional capabilities can be added easily. The current version has some limitations, however, in the area of of security and usability for computing in a distributed environment. Because the potential solutions to these problems are related, we discuss them together.

In any networked or distributed computing system, two important questions must be answered: How does the system protect itself from malicious or careless users (and these users from each other), and how does it present itself to the users? The first issue is one of security; the second is one of ease of use. NEOS currently has only rudimentary security features, and the interfaces do not allow access from a user program. A more traditional but more flexible and powerful method for accessing numerical capabilities is the procedural (subroutine) interface. The network analogue to this interface, the remote procedure call (RPC) [6], is useful for some parts of the security question as well. To better understand the security issues, we first outline the problem.

To make the discussion concrete, we consider the case of determining a minimizer of a function $f : \mathbb{R}^n \mapsto \mathbb{R}$. The most precise description of this problem is often the code that implements $f$. This code most likely contains loops and array accesses, and possibly function calls. If the NEOS Server simply accepts arbitrary code to describe the optimization problem, the systems running NEOS are subject to some security risks.

The most common security risk occurs if the user is not trusted. Lack of trust does not mean that the user is malicious, just that the work they send cannot be trusted not to cause problems. In this case we identify two issues: rogue programs and denial-of-service attacks.

Rogue programs are a security issue because the description of the optimization problem may crash the Server or even compromise the system. For problems (for example, linear programming or distance geometry) that are described by relatively simple data, it is easy to check that the data is valid. For programs, such checking is much harder. Our current system applies a number of techniques to reduce the chance of harm to the system. For example, the NEOS Server runs as a separate user with no special privileges. However, since the user can submit Fortran or C programs, it is possible that a programming error or a deliberate subterfuge could harm NEOS or, through some unplugged security problem, compromise the entire system. Solutions to this problem range from provably safe problem descriptions (either a specialized modeling language or strongly typed languages such as Java) to provably safe interfaces, with user code and NEOS code running on different systems.

Another security issue arises, even with well-meaning users, when programs use large amounts of computing—more than their fair share. When this is done maliciously, it is called a denial-of-service attack. The amount of time used by a single request is relatively easily controlled, and most scheduling systems provide this service. Harder to contain are floods of requests, each generating a separate request. The NEOS Server may need to maintain records of use as well as interacting with the scheduling system to ensure that the Server does not make excessive demands on the system.

A security risk also occurs if the user is trusted but the connection is not, because there may be eavesdroppers. This requires merely that some secure transaction service be used. Such services are becoming commonplace on the Web (for example, for credit card transactions), though usually for small amounts of data.

Providing both a more secure execution environment for NEOS and a more flexible interface makes use of the same technique: the remote procedure call. Conventionally, programmers of scientific applications call library routines from their Fortran or C program to perform some service, such as minimizing a function. For example, the user may call a procedure of the form

```
call uco( fcn, x, ... )
```

to find a minimizer of $f$. A remote procedure call looks just like a regular procedure call to the user; the implementation, however, makes contact with a remote server (NEOS in our case) and passes to the server the arguments of the call. The implementation may look like

```
call neos_uco( 'fcn.f', x, ... )
```

where `'fcn.f'` is the name of the file containing the code that defines the function $f$.

A remote procedure call interface would be a significant improvement over the interface provided by the NEOS Submission tool because such an interface would make it possible to incorporate NEOS capabilities into existing code. The remote procedure call interface is a relatively simple addition to the existing NEOS services that is secure and easy to use for those optimization problems that are completely described by simple data, such as linear programming problems. The data can be checked for validity by the server; encryption can be used if necessary to keep the data safe from eavesdroppers. If the data is Fortran or C code, however, checking for validity is far more difficult. If the user is not trusted, we can use a reverse communication approach or a provably safe form of code.

In the reverse communication approach, the user's function is executed on the user's machine. The RPC mechanism is used to request information from the user and to return information to the Server. For example, the user could be requested to evaluate the function $f$ at the current iterate $x$ and return $f(x)$ to the Server. This maintains the security against rogue code by executing the user's code on the user's machine and the optimization code on the NEOS-provided machine.

A drawback of this approach is that the cost of each of these remote procedure calls can be large. Just the delay from speed-of-light propagation can amount to milliseconds in a widely distributed network. In view of this limitation, this approach would be suitable only for problems in which function evaluations require considerable computing power so that the cost of the remote procedure call is insignificant.

The option to use a provably safe form of the code has attracted attention because of Java, a language related to C++ that contains a number of features and restrictions to allow programs to be executed without fearing that the system running them could be compromised. Many of the restrictions are irrelevant to most numerical code; for example, Java has no function pointers and restricts the kinds of routines (other than those provided by the user) that can be called.

In the most obvious approach, the user writes code in Java instead of Fortran or C, and the Java code is uploaded into the NEOS Server by the remote procedure call. There is another option; the optimization code provided by NEOS could be downloaded directly into the user's running application (as long as that application is running in a Java environment). The interface would be the same; only the source of the cycles used to run the optimization code would be different. In this model, the procedure interface is a sort of automatic download-and-install operation that would be suitable only for small problems, and would not provide access to large-scale computing resources.

Java is not the only choice; a suitable subset of Fortran could be constructed by restricting some features of the language and performing strict compile and runtime checking. Such a network-safe version of Fortran could provide easy and reliable access to distributed services to existing programs.

# Acknowledgments

The ideas discussed in this paper have evolved during the development of the NEOS Server. Many have contributed to this effort, but Michael Mesnier deserves special credit. Other researchers that have contributed to these ideas are Joe Czyzyk, Michael Ferris, Jorge Nocedal and Steve Wright.

# References

[1] Averick, B.M., R.G. Carter, J.J. Moré and G.-L. Xue (1992) "The MINPACK-2 test problem collection", Preprint MCS-P153-0694, Mathematics and Computer Science Division, Argonne National Laboratory, Argonne, IL.

[2] Bischof, C., A. Bouaricha, P. Khademi and J.J. Moré (1995), "Computing gradients in large-scale optimization using automatic differentiation", Preprint MCS-P488-0195, Argonne National Laboratory, Argonne, IL. To appear in *INFORMS Journal on Computing.*

[3] Bischof, C., A. Carle and P. Khademi (1994), "Fortran 77 interface specification to the SparsLinC library", Technical Report ANL/MCS-TM-196, Argonne National Laboratory, Argonne, IL.

[4] Bischof, C., A. Carle, P. Khademi and A. Mauer (1994), "The ADIFOR 2.0 system for the automatic differentiation of Fortran 77 programs", Preprint MCS-P381-1194, Argonne National Laboratory, Argonne, IL. Also available as CRPC-TR94491, Center for Research on Parallel Computation, Rice University, Houston, TX.

[5] Blaney, J.M. and J.S. Dixon (1994), "Distance geometry in molecular modeling", in *Reviews in Computational Chemistry* **5** (K.B. Lipkowitz and D.B. Boyd, eds), VCH Publishers (New York), 299–335.

[6] Bloomer, J. (1992), *Power Programming with RPC,* O'Reilly & Associates, Inc. (Sebastopol, CA).

[7] Bouaricha, A. and J.J. Moré (1995), "Impact of partial separability on large-scale optimization", Preprint MCS-P487-0195, Argonne National Laboratory, Argonne, IL. To appear in *Computational Optimization and Applications.*

[8] Brünger, A.T. and M. Nilges (1993), "Computational challenges for macromolecular structure determination by X-ray crystallography and solution NMR-spectroscopy", *Q. Rev. Biophys.* **26**, 49–125.

[9] Conn, A.R., N.I.M. Gould and P.L. Toint (1992), *LANCELOT,* Springer Series in Computational Mathematics, Springer-Verlag (Berlin).

[10] Crippen, G.M. and T.F. Havel (1988), *Distance Geometry and Molecular Conformation,* John Wiley & Sons (New York).

[11] Czyzyk, J., M.P. Mesnier and J.J. Moré (1996), *The Network-Enabled Optimization System (NEOS) Server*, Preprint MCS-P615-0996, Argonne National Laboratory, Argonne, IL.

[12] Ferris, M.C., M.P. Mesnier and J.J. Moré (1996), "The NEOS Server for complementarity problems: PATH", Technical Report 96-08, University of Wisconsin, Madison, WI. Also available as MCS-P616-1096, Mathematics and Computer Science Division, Argonne National Laboratory, Argonne, IL.

[13] Foster, I. and C. Kesselman (1998), "Globus: A metacomputing infrastructure toolkit". To appear in *International J. Supercomputer Applics.*

[14] Goodman, J., R. Kohn and L. Reyna (1986), "Numerical study of a relaxed variational problem from optimal design", *Comput. Methods Appl. Mech. Engrg* **57**, 107–127.

[15] Griewank, A., D. Juedes and J. Utke (1996), "ADOL-C: A package for the automatic differentiation of algorithms written in C/C++", *ACM Trans. Math. Software* **22**, 131–167.

[16] Grinshaw, A. and W. Wolf (1996), "Legion—A view from 50,000 feet", in *Proceedings of the 5th IEEE Symposium on High Performance Distributed Computing,* IEEE Computer Society Press (Los Alamitos, CA), 89–99.

[17] Havel, T.F. (1991), "An evaluation of computational strategies for use in the determination of protein structure from distance geometry constraints obtained by nuclear magnetic resonance", *Prog. Biophys. Mol. Biol.* **56**, 43–78.

[18] Havel, T.F. (1995), "Distance geometry", in *Encyclopedia of Nuclear Magnetic Resonance* (D.M. Grant and R.K. Harris, eds), John Wiley & Sons (New York), 1701–1710.

[19] Kuntz, I.D., J.F. Thomason and C.M. Oshiro (1993), "Distance geometry", in *Methods in Enzymology* **177** (N.J. Oppenheimer and T.L. James, eds), Academic Press (New York), 159–204.

[20] Litzkow, M.J., M. Livny and M.W. Mutka (1988), "Condor—A hunter of idle workstations", in *Proceedings of the 8th International Conference on Distributed Computing Systems,* IEEE Computer Society Press (Washington, DC), 108–111.

[21] Moré, J.J. and Z. Wu (1995), "$\varepsilon$-optimal solutions to distance geometry problems via global continuation", in *Global Minimization of Nonconvex Energy Functions: Molecular Conformation and Protein Folding* (P.M. Pardalos, D. Shalloway and G. Xue, eds), American Math. Soc. (Providence, RI), 151–168.

[22] Moré, J.J. and Z. Wu (1996), "Distance geometry optimization for protein structures", Preprint MCS-P628-1296, Argonne National Laboratory, Argonne, IL.

[23] Toint P.L. (1986), "Numerical solution of large sets of algebraic nonlinear equations", *Math. Comp.* **46**, 175–189.

[24] Toint P.L. (1987), "On large scale nonlinear least squares calculations", *SIAM J. Sci. Statist. Comput.* **8**, 416–435.

[25] Toint, P.L. and D. Tuyttens (1990), "On large-scale nonlinear network optimization", *Math. Programming* **48**, 125–159.

[26] Toint, P.L. and D. Tuyttens (1992), "LSNNO: A Fortran subroutine for solving large-scale nonlinear network optimization problems", *ACM Trans. Math. Software* **18**, 308–328.

[27] Torda, A.E. and W.F. van Gunsteren (1992), "Molecular modeling using nuclear magnetic resonance data", in *Reviews in Computational Chemistry* **3** (K.B. Lipkowitz and D.B. Boyd, eds), VCH Publishers (New York), 143–172.

[28] Wall, L., T. Christiansen and R.L. Schwartz (1996), *Programming Perl,* O'Reilly & Associates, Inc., 2nd edn (Sebastopol, CA).

# New Versions of qd for Products of Bidiagonals

*Beresford N. Parlett*

Mathematics Department and Computer Science Division, EECS Department, University of California, Berkeley, CA 94720, USA[1]

**Abstract**

Tridiagonal matrices are a popular condensed form of general square matrices. This paper argues that it is preferable to represent a tridiagonal matrix as a product of two bidiagonal matrices. In the eigenvalue context a translate of a given matrix, if needed, always permits triangular factorization. The qd-algorithm of H. Rutishauser is, at an abstract level, equivalent to the LR algorithm, another invention of Rutishauser. New, and more accurate versions of qd are introduced. It is shown that these new algorithms applied to bidiagonal matrices $L$ and $U$ are equivalent to applying the (two-sided) Gram–Schmidt process to the columns of $L$ and the rows of $U$. In this sense they are independent of the original qd algorithm.

Some results on accuracy are quoted. This paper is a reduced version of [5].

## 1 Introduction

Let us think about ways to find eigenvalues of tridiagonal matrices. An important special case is to compute singular values of bidiagonal matrices. The discussion is addressed both to specialists in matrix computation and to other scientists. The reason for hoping to communicate with two such diverse sets of readers at the same time is that the content of the survey, though of recent origin, is quite elementary and does not demand familiarity with much beyond triangular factorization and the Gram–Schmidt process for orthogonalizing a set of vectors. For some readers the survey will cover familiar territory but from a novel perspective. The justification for presenting these ideas is that they lead to new variations of current methods that run a lot faster while achieving greater accuracy.

Tridiagonal matrices have received a great deal of attention since the 1950s. The $(i, j)$ entries of these matrices vanish if $|i-j| > 1$ and the interest in them stems from the fact that they are the most narrowly banded similarity class of a matrix that can be obtained by a finite number of similarity transformations using rational expressions in the matrix entries.

[1]Supported by ONR, Contract N000014-90-J-1372.

It should be mentioned that the tridiagonal form is probably also too condensed for the most difficult cases [4], but it is rich enough to suffice for many applications and we shall stay with it here.

Our topic, new versions of qd algorithms, will be developed as the consequence of two ideas.

The first concerns the representation of tridiagonal matrices and we mention it briefly here. In the eigenvalue context there is no loss of generality in supposing that our tridiagonals are normalized so that all entries in positions $(i, i+1)$ are either 0 or 1. Zeros here make calculations easier so we may assume that all these entries are 1. Such tridiagonals are denoted by $J$. Most, but not all, such $J$ permit triangular factorization

$$J = LU,$$

where the precise forms of $L$ and $U$ are shown at the beginning of the next section. It is clear that in the $n \times n$ case $L$ and $U$ together are defined by $2n-1$ parameters; exactly the same degree of freedom as in $J$. Section 3 argues that the pair $L$, $U$ is preferable to $J$ itself. Consequently all transformations on tridiagonals should be re-examined in this representation.

The second idea relates to the LR algorithm (LR) discovered by H. Rutishauser in 1957 [7], and presented in Section 5 here. When LR is rewritten in the $L, U$ representation one obtains the (progressive) qd-algorithm. The letters q and d are lower case because they do not stand for matrices and the matrix computation community tries to reserve capital letters for matrices. Thus q here has nothing to do with the Q of the QR factorization.

In the new representation the LR algorithm spends most of its time computing the triangular factorization not of a single matrix, but of a product, namely $UL$. It is not well enough appreciated that finding the $LU$ factorization of any product $BC$ is equivalent to applying a generalized Gram–Schmidt process to the rows of $B$ and the columns of $C$ so that $B = \hat{L}P^*$, $C = Q\hat{U}$, and $P^*Q$ is diagonal. When this Gram–Schmidt process is applied to $UL$ in an efficient manner one obtains a little known variant of qd, called the differential qd algorithm (dqd), that requires a little more arithmetic effort than qd itself. Rutishauser discovered dqd [9] near the end of his life and never mentioned the shifted version dqds that K.V. Fernando and I discovered [2] independently of Rutishauser's work, in 1991, while trying to improve on the Demmel and Kahan QR algorithm [1] for computing singular values of bidiagonals. The connection of dqd with the generalized Gram–Schmidt process on bidiagonals is new and constitutes the second idea that underpins this survey.

The presentation here runs completely counter to history. The paper [2] develops dqds in the singular value context, gives historical comments and shows the connections with continued fractions. However, none of that is necessary and it is simplicity we pursue here.

This survey develops several qd-algorithms (six in all) in a matrix context in the most elementary way. It is not difficult to see several directions in which these ideas may be generalized or modified to good effect.

The differential forms of qd algorithms are the right ones for parallel computation.

A sceptic might say that since no one uses LR algorithms there is no point in finding fancy versions of them. In the general case there is still much work to be done in finding clever shift strategies that approach an eigenvalue in a stable way. However, in the symmetric case even the current simple shift strategies achieve high relative accuracy in all eigenvalues and are between 2 and 10 times faster than QR [2]. Yet the most powerful argument in favour of qd algorithms may turn out to be the efficient computation of accurate eigenvectors.

# 2 Bidiagonals versus tridiagonals

Bidiagonal matrices of a special form will play a leading role in this essay. Whenever possible $6 \times 6$ matrices will be used to illustrate the general pattern.

$$L = \begin{bmatrix} 1 & & & & & \\ l_1 & 1 & & & & \\ & l_2 & 1 & & & \\ & & l_3 & 1 & & \\ & & & l_4 & 1 & \\ & & & & l_5 & 1 \end{bmatrix}, \quad U = \begin{bmatrix} u_1 & 1 & & & & \\ & u_2 & 1 & & & \\ & & u_3 & 1 & & \\ & & & u_4 & 1 & \\ & & & & u_5 & 1 \\ & & & & & u_6 \end{bmatrix}.$$

To save space these matrices may be written as

$$\begin{aligned} L &= bidiag\begin{pmatrix} 1 & & 1 & & 1 & & 1 & & 1 & & 1 \\ & l_1 & & l_2 & & l_3 & & l_4 & & l_5 & \end{pmatrix}, \\ U &= bidiag\begin{pmatrix} & 1 & & 1 & & 1 & & 1 & & 1 & \\ u_1 & & u_2 & & u_3 & & u_4 & & u_5 & & u_6 \end{pmatrix}. \end{aligned}$$

The pair $L, U$ determine two tridiagonal matrices; first

$$J = LU = \begin{bmatrix} u_1 & 1 & & & & \\ l_1u_1 & l_1+u_2 & 1 & & & \\ & l_2u_2 & l_2+u_3 & 1 & & \\ & & l_3u_3 & l_3+u_4 & 1 & \\ & & & l_4u_4 & l_4+u_5 & 1 \\ & & & & l_5u_5 & l_5+u_6 \end{bmatrix},$$

which may be written

$$J = tridiag\begin{pmatrix} & 1 & & 1 & \bullet & 1 & \\ u_1 & & l_1+u_2 & & \bullet \quad \bullet & & l_5+u_6 \\ & l_1u_1 & & l_2u_2 & \bullet & l_5u_5 & \end{pmatrix},$$

and second

$$J' = UL = \begin{bmatrix} u_1 + l_1 & 1 & & & & \\ u_2 l_1 & u_2 + l_2 & 1 & & & \\ & u_3 l_2 & u_3 + l_3 & 1 & & \\ & & u_4 l_3 & u_4 + l_4 & 1 & \\ & & & u_5 l_4 & u_5 + l_5 & 1 \\ & & & & u_6 l_5 & u_6 \end{bmatrix},$$

which may be written

$$J' = tridiag\left( \begin{array}{ccccccc} & 1 & & 1 & \bullet & 1 & \\ u_1 + l_1 & & u_2 + l_2 & & \bullet \quad \bullet & & u_6 \\ & u_2 l_1 & & u_3 l_2 & \bullet & u_6 l_5 & \end{array} \right).$$

Note that both tridiagonals have their superdiagonal entries, that is, entries $(j, j+1)$, equal to 1. Also note that $J' = L^{-1}JL$. The reader should note the pattern of indices in $J$ and $J'$ because frequent reference will be made to them throughout the survey.

The attractive feature here is that because the 1's need not be represented explicitly the factored form of $J$ requires no more storage than $J$ itself; there are $2n - 1$ parameters for $n \times n$ matrices in each case.

**Advantages of the factored form**

1. $L, U$ determine the entries of $J$ to greater than working precision because the addition and multiplication of $l$'s and $u$'s is implicit. Thus $J_{ii}$ is given by $l_{i-1} + u_i$ implicitly but by $\mathrm{fl}(l_{i-1} + u_i)$ explicitly.

2. The mapping $L, U \to J$ is naturally parallel; for example $l * u$ gives the off-diagonal entries. In contrast the mapping $J \to L, U$, that is Gaussian elimination, is intrinsically sequential.

3. Singularity of $J$ is detectable by inspection when $L$ and $U$ are given, but only by calculation from $J$.

4. Solution of $Jx = b$ takes half the time when $L$ and $U$ are available.

5. In many cases the entries in $L$ and $U$ are better than those of the product $LU$ for defining the eigenvalues of $LU$ to high accuracy.

**Disadvantages of the factored form**

The mapping $J \to L, U$ is not everywhere defined. Even when the factorization exists it can happen that $\|L\|$ and $\|U\|$ greatly exceed $\|J\|$. This is very bad for applying the LR algorithm but harmless when eigenvectors are to be calculated. So we should be careful to consider the goal before stigmatizing a process as unstable. Moreover in the eigenvalue context we are free to replace $J$ by $J - \sigma I = LU$ for some suitably chosen shift $\sigma$ that gives acceptable $L$ and $U$.

Frequently we make the *nonzero assumption*: $l_j u_j \neq 0,\ j = 1, \ldots, n-1$. If $u_n = 0$ then a glance at $J' = UL$ shows that its last row is zero. However, it is not valid to simply discard $l_{n-1}$ along with $u_n$. In other words we do not have a factored form of the leading $(n-1) \times (n-1)$ principal submatrix of $J'$ unless $l_{n-1}$ is negligible compared with $u_{n-1}$. Similarly if $u_1$ is zero we must not ignore $l_1$. In fact any zero values among $\{l_j,\ u_j;\ j = 1, \ldots, n-1\}$ are readily exploited.
*Splitting*: If $l_j = 0,\ j < n$, then the spectrum of $J$ is the union of the spectra of two smaller tridiagonals given in factored form by $\{l_i,\ u_i;\ i = 1, j\}$ and $\{l_i,\ u_i;\ i = j+1, n\}$. Here $l_n = 0$.
*Singularity*: If $u_j = 0,\ j \leq n$, then zero is an eigenvalue. However, some computation is necessary to deflate this eigenvalue and obtain $L$ and $U$ factors of an $(n-1) \times (n-1)$ matrix. One pass of the qd algorithm (given later) will suffice in exact arithmetic.

Any tridiagonal matrix that does not split, or its transpose, is diagonally similar to a form with 1's above the diagonal, that is, a $J$ matrix. So for eigenvalue hunting there is no loss of generality in using this normalization.

In the past most attention has been paid to the *positive case*; $l_i > 0$, $i = 1, \ldots, n-1,\ u_j > 0,\ j = 1, \ldots, n$. Note in passing the following standard results.

**Lemma 1** *If $l_i u_i > 0,\ i = 1, \ldots, n-1$, then $J$ is symmetrizable by a diagonal similarity and the number of positive (negative) $u_i$ is the number of positive (negative) eigenvalues.*

**Lemma 2** *If $l_i u_{i+1} > 0,\ i = 1, \ldots, n-1$, then $J'$ is symmetrizable by a diagonal similarity and the number of positive (negative) $u_i$ is the number of positive (negative) eigenvalues.*

For a real symmetric or complex Hermitian matrix a preliminary reduction to tridiagonal form has proved to be a stable and efficient step in computing the eigenvalues. In the general case preliminary reduction to tridiagonal form has been less successful. Stability is not guaranteed for any current methods

[4]. Sometimes users are lucky but as larger and larger matrices are tried unsatisfactory experiences are more frequent. It may well be that the tridiagonal form is too compact for the difficult cases. The attentive reader will note in the following pages that some of the algorithms can be extended to fatter forms, such as block tridiagonal, but such ideas will not be pursued here.

# 3 Stationary qd algorithms

Triangular factors change in a complicated way under translation. Given $L$ and $U$ of the form given in Section 2 the task here is to compute $\bar{L}$ and $\bar{U}$ so that

$$J - \sigma I = LU - \sigma I = \bar{L}\bar{U}$$

for a given suitable shift $\sigma$. Equating entries on each side shows that

$$\begin{aligned} l_i + u_{i+1} - \sigma &= \bar{l}_i + \bar{u}_{i+1}, \quad i = 0, \ldots, n-1, \; l_0 = 0, \\ l_i u_i &= \bar{l}_i \bar{u}_i, \quad i = 1, \ldots, n-1. \end{aligned}$$

These relations yield the so called stationary qd algorithm:

*stqd*($\sigma$): $\bar{u}_1 = u_1 - \sigma$;
  **for** $i = 1, n-1$
    $\bar{l}_i = l_i u_i / \bar{u}_i$
    $\bar{u}_{i+1} = l_i + u_{i+1} - \sigma - \bar{l}_i$
  **end for**.

Naturally it fails if $\bar{u}_i = 0$ for some $i < n$.

At this point the sceptical reader might object that stqd($\sigma$) is exactly the same algorithm that would be obtained by forming $J - \sigma I$ and performing Gaussian elimination. Indeed if stqd($\sigma$) is executed with the operations proceeding from left to right, for example,

$$\bar{u}_{i+1} = \mathrm{fl}(\mathrm{fl}(\mathrm{fl}(l_i + u_{i+1}) - \sigma) - \bar{l}_i),$$

then the two procedures are not just mathematically equivalent but also computationally identical. However, it is not necessary to follow this left-to-right ordering. For example, one could write

$$\bar{u}_{i+1} = (l_i - \bar{l}_i - \sigma) + u_{i+1}$$

and, if the compiler respects parentheses, then stqd($\sigma$) will quite often produce different output than Gaussian elimination on $J - \sigma I$. If $l_i$ and $\bar{l}_i$ are much larger than $u_{i+1}$ then the second form is more accurate than the first.

The preceding thoughts lead to an alternative algorithm, easily missed, for $\bar{L}$ and $\bar{U}$. It involves more arithmetic effort and an auxiliary storage cell but has some striking advantages in accuracy for finite precision arithmetic.

To derive the algorithm define variables $\{t_i\}$ by

$$t_{i+1} \equiv \bar{u}_{i+1} - u_{i+1} = l_i - \bar{l}_i - \sigma.$$

Observe that

$$\begin{aligned} t_{i+1} &= l_i - l_i u_i / \bar{u}_i - \sigma \\ &= l_i(\bar{u}_i - u_i)/\bar{u}_i - \sigma \\ &= t_i l_i / \bar{u}_i - \sigma. \end{aligned}$$

For reasons that are not clear Rutishauser called the associated algorithm the *differential* form of stqd. We call it dstqd.

dstqd($\sigma$): $t_1 = -\sigma$;
**for** $i = 1, n-1$
$\bar{u}_i = u_i + t_i$
$\bar{l}_i = u_i(l_i/\bar{u}_i)$
$t_{i+1} = t_i(l_i/\bar{u}_i) - \sigma$
**end for**
$\bar{u}_n = u_n + t_n$.

In practice the $t$-values may be written over each other in a single variable $t$. If the common subexpression $l_i/\bar{u}_i$ is recognized then only one division is needed. Thus dstqd exchanges a subtraction for a multiplication so the extra cost is not excessive.

At first sight stqd may not seem relevant to the eigenvalue problem but if $\lambda$ is a very accurate approximation to an eigenvalue then dstqd($\lambda$) is needed for inverse iteration. In this application huge values among the $\{\bar{u}_i\}$ are to be expected and do not have a deleterious effect on the computed eigenvector. In fact dstqd($\sigma$) may be used to find eigenvalues too by extracting a good approximate eigenvalue from the $t$-values for a shift.

# 4 Progressive qd algorithms

This section seeks the triangular factorization of $J' - \sigma I$, not $J - \sigma I$:

$$J' - \sigma I = UL - \sigma I = \hat{L}\hat{U}$$

for a suitable shift $\sigma$. Equating entries on each side of the defining equation gives the so called rhombus rules of H. Rutishauser, see [6] in German and, in English, [3].

$$u_{i+1} + l_{i+1} - \sigma = \hat{l}_i + \hat{u}_{i+1} \quad \text{and} \quad l_i u_{i+1} = \hat{l}_i \hat{u}_i.$$

These relations give the so called progressive qd algorithm with shift which we call qds($\sigma$).

$$
\begin{aligned}
\text{qds}(\sigma)\colon\ & \hat{u}_1 = u_1 + l_1 - \sigma; \\
& \textbf{for } i = 1, n-1 \\
& \qquad \hat{l}_i = l_i u_{i+1}/\hat{u}_i \\
& \qquad \hat{u}_{i+1} = u_{i+1} + l_{i+1} - \sigma - \hat{l}_i \\
& \textbf{end for}.
\end{aligned}
$$

The algorithm qds fails when $\hat{u}_i = 0$ for some $i < n$. In contrast to the stationary algorithm the mapping $\sigma, L, U \to \hat{L}, \hat{U}$ is nontrivial even when $\sigma = 0$. When $\sigma = 0$ we write simply qd, not qds.

At this point the sceptical reader might object that qds($\sigma$) is exactly the same algorithm that would be obtained by forming $J' - \sigma I$ and performing Gaussian elimination. Indeed if the operations are done proceeding from left to right, for example,

$$\hat{u}_{i+1} = \mathrm{fl}(\mathrm{fl}(\mathrm{fl}(u_{i+1} + l_{i+1}) - \sigma) - \hat{l}_i),$$

then the two procedures are not just mathematically equivalent but also computationally identical. However, it is not necessary to follow this ordering. For example, one could write

$$\hat{u}_{i+1} = (l_{i+1} - \hat{l}_i - \sigma) + u_{i+1}$$

and, if the compiler respects parentheses, then the output will quite often be different.

There is an alternative implementation of qds that is easy to miss. In fact Rutishauser never wrote it down. The new version is slightly slower than qds but has compensating advantages. Here we derive it simply as a clever observation, leaving to later sections the task of making it independent of qds.

As suggested in an earlier paragraph we might define an auxiliary variable

$$d_{i+1} \equiv u_{i+1} - \hat{l}_i - \sigma \ (= \hat{u}_{i+1} - l_{i+1}).$$

Observe that

$$
\begin{aligned}
d_{i+1} &= u_{i+1} - (l_i u_{i+1}/\hat{u}_i) - \sigma \\
&= (u_{i+1}(\hat{u}_i - l_i)/\hat{u}_i) - \sigma \\
&= (u_{i+1} d_i/\hat{u}_i) - \sigma.
\end{aligned}
$$

Rutishauser seems to have discovered the unshifted version two or three years before he died, perhaps 15 years after discovering qd, but he did not make much use of it [9]. He called it the *differential* qd algorithm (dqd for short) and so we call the new shifted algorithm dqds (*differential qd* with *shifts*).

dqds($\sigma$): $d_1 = u_1 - \sigma$;
**for** $i = 1, n-1$
$\hat{u}_i = d_i + l_i$
$\hat{l}_i = l_i(u_{i+1}/\hat{u}_i)$
$d_{i+1} = d_i(u_{i+1}/\hat{u}_i) - \sigma$
**end for**
$\hat{u}_n = d_n$.

By definition, dqd = dqds(0). In the positive case dqd requires *no subtractions* and enjoys very high relative stability, see Section 8. In practice each $d_{i+1}$ may be written over its predecessor in a single variable $d$. Looking ahead we mention that the quantity $\min |d_j|$ gives useful information on the eigenvalue nearest 0.

# 5 The LR algorithm for $J$ matrices

As mentioned in Section 1 our exposition reverses the historical process. Rutishauser discovered LR by interpreting qd in terms of bidiagonal matrices, a brilliant and fruitful insight. This is worth explaining. By definition, the LR transform of $J$ is $J'$ and of $J'$ is the matrix $J''$ defined in two steps by

$$J' = \hat{L}\hat{U}, \quad J'' = \hat{U}\hat{L}.$$

Now qd applied to $L$ and $U$ yields $\hat{L}$ and $\hat{U}$ and so defines $J''$ implicitly. There is no need to form $J'$ or $J''$.

When shifts are employed the situation is a little more complicated. It is necessary to look at two successive steps with shifts $\sigma_1$ and $\sigma_2$.

In shifted LR

$$\begin{aligned} J_1 - \sigma_1 I &= L_1 U_1, \\ J_2 &= U_1 L_1 + \sigma_1 I, \\ J_2 - \sigma_2 I &= L_2 U_2, \\ J_3 &= U_2 L_2 + \sigma_2 I. \end{aligned}$$

In other words, the shifts are restored so that $J_1$, $J_2$, $J_3$ are similar. Note that

$$\begin{aligned} J_2 = U_1 L_1 + \sigma_1 I &= L_1^{-1}(J_1 - \sigma_1 I)L_1 + \sigma_1 I \\ &= L_1^{-1} J_1 L_1. \end{aligned}$$

However, if $J_2$ is not to be formed one cannot *explicitly* add $\sigma_1$ back to the diagonal. On the other hand

$$J_2 - \sigma_2 I = U_1 L_1 - (\sigma_2 - \sigma_1 I) = L_2 U_2.$$

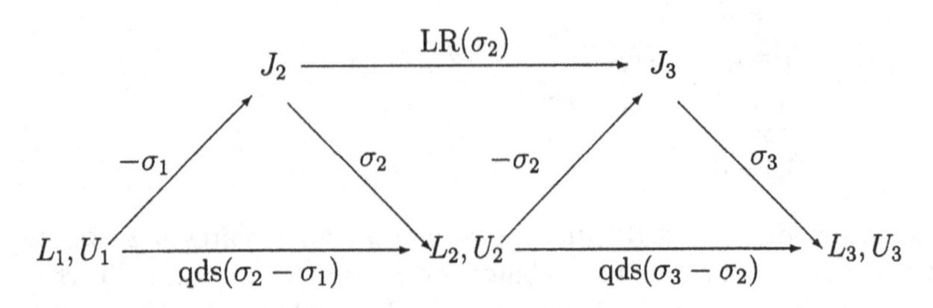


**Figure 5.1** Relation of LR to qds.

Thus to find $L_2$ and $U_2$ from $L_1$ and $U_1$ it is only necessary to apply qds($\sigma_2 - \sigma_1$). In other words, to get qds equivalent to LR with shifts $\{\sigma_i\}_{i=1}^{\infty}$ it is necessary to use the differences $(\sigma_i - \sigma_{i-1})$ with qds. In LR the shifts should converge to an eigenvalue of the original $J$ or $J'$. In qds the shifts should converge to 0 and $u_n \to 0$, $l_{n-1} \to 0$ too and all shifts must be accumulated. It is worth recording the relationship in a diagram in Figure 5.1.

In practice the LR algorithm avoids explicit calculation of the $L$'s and $U$'s and the transformation $J_i \to J_{i+1}$ is effected via a sequence of elementary similarity transformations. As implemented in the late 1950s and early 1960s, LR proved insufficiently reliable and was displaced by the QR algorithm in the mid 1960s. In one important class of applications both LR and qds were accurate and efficient: the positive case $l_i > 0$, $u_i > 0$ for all $i$.

For comparison purposes we present two implementations of LR: the explicit and the implicit shift versions. Let

$$J = tridiag \begin{pmatrix} & 1 & & 1 & \bullet & & 1 & \\ \alpha_1 & & \alpha_2 & & \bullet & \bullet & & \alpha_n \\ & \beta_1 & & \beta_2 & \bullet & & \beta_{n-1} & \end{pmatrix},$$

let $J - \sigma I = LU$, $\hat{J} = UL + \sigma I$ and write $\hat{J}$ in the same notation as $J$. The matrix $L$ may be written as a product of lower triangular plane transformers $N_i$. So

$$U = N_{n-1}^{-1} \cdots N_1^{-1}(J - \sigma I), \quad \hat{J} = UN_1 \cdots N_{n-1} + \sigma I$$

and the active part of $N_i$ is $\begin{pmatrix} 1 & 0 \\ e_i & 1 \end{pmatrix}$, where the multiplier $e_i$ is in position $(i+1, i)$. The following diagram shows a typical stage.

$$\begin{pmatrix} 1 & 0 \\ -e_j & 1 \end{pmatrix} \begin{pmatrix} d_j e_{j-1} & d_j & 1 \\ \beta_j e_{j-1} & \beta_j & \alpha_{j+1} - \sigma \end{pmatrix} = \begin{pmatrix} \hat{\beta}_{j-1} & d_j & 1 \\ 0 & 0 & d_{j+1} \end{pmatrix},$$

$$\begin{pmatrix} d_j & 1 \\ 0 & d_{j+1} \\ 0 & \beta_{j+1} \end{pmatrix} \begin{pmatrix} 1 & 0 \\ e_j & 1 \end{pmatrix} = \begin{pmatrix} d_j + e_j & 1 \\ e_j d_{j+1} & d_{j+1} \\ e_j \beta_{j+1} & \beta_{j+1} \end{pmatrix}.$$

LR (explicit shift $\sigma$): $d_1 = \alpha_1 - \sigma$;
**for** $i = 1, n-1$
  **if** $d_i = 0$ **then** exit (fail)
  $e_i = \beta_i / d_i$
  $\hat{\alpha}_i = d_i + (e_i + \sigma)$
  $d_{i+1} = \alpha_{i+1} - (e_i + \sigma)$
  $\hat{\beta}_i = d_{i+1} * e_i$
**end for**
$\alpha_n = d_n + \sigma$.

In practice we write $d$ and $e$ for $d_i$ and $e_i$.

To derive the implicit shift version we note first that

$$\hat{J} = N_{n-1}^{-1} \cdots N_1^{-1} J N_1 \cdots N_{n-1}$$

and, of more importance, that the $2 \times 2$ submatrix

$$\begin{pmatrix} \hat{\beta}_{j-1} & d_j \\ \beta_j e_{j-1} & \beta_j \end{pmatrix}$$

has rank one in exact arithmetic. Thus the multiplier $e_j$ could be computed from $e_j = \beta_j e_{j-1} / \hat{\beta}_{j-1}$ and then the shift $\sigma$ disappears from the inner loop.

The initial value $e_1 = \beta_1 / (\alpha_1 - \sigma)$ is the only occasion on which $\sigma$ appears. Indeed if $\alpha_1 \gg \sigma$ then much of the information in $\sigma$ is irretrievably lost.

The algorithm is sometimes described as chasing the bulge $\beta_j e_{j-1}$ in position $(j+1, j-1)$ down the matrix and off the bottom as $j = 2, 3, \ldots, n-1$ and $n$. We write $\delta$ instead of $d$ to emphasize that these quantities differ from the corresponding ones in the explicit shift algorithm.

LR (implicit shift $\sigma$): $\delta = \alpha_1$
**if** $\delta = \sigma$ **then** exit (fail)
$e = \beta_1 / (\delta - \sigma)$
**for** $i = 1, n-1$
  $\hat{\alpha}_i = \delta + e$
  $\hat{\beta}_i = \beta_i - e * (\hat{\alpha}_i - \alpha_{i+1})$
  $\delta = \alpha_{i+1} - e$
  **if** $\hat{\beta}_i = 0$ **then** exit (fail)
  $e = e * \beta_{i+1} / \hat{\beta}_i$
**end for**
$\hat{\alpha}_n = \delta$.

The attraction of this algorithm is that it employs nothing but explicit similarities on $J$.

# 6 Gram–Schmidt factors

This section shows that *dqd* could have been discovered independently of *qd*.

To most people the Gram–Schmidt process is the standard way of producing an orthonormal set of vectors $\boldsymbol{q}_1, \boldsymbol{q}_2, \ldots, \boldsymbol{q}_k$ from a linearly independent set $\boldsymbol{f}_1, \boldsymbol{f}_2, \ldots, \boldsymbol{f}_k$. The defining property is that $span(\boldsymbol{q}_1, \boldsymbol{q}_2, \ldots, \boldsymbol{q}_j) = span(\boldsymbol{f}_1, \boldsymbol{f}_2, \ldots, \boldsymbol{f}_j)$, for each $j = 1, 2, \ldots, k$. The matrix formulation of this process is the QR factorization: $F = QR$, where $F = [\boldsymbol{f}_1, \boldsymbol{f}_2, \ldots, \boldsymbol{f}_k]$, $Q = [\boldsymbol{q}_1, \boldsymbol{q}_2, \ldots, \boldsymbol{q}_k]$ and $R$ is $k \times k$ and upper triangular.

The generalization of this process to a pair of vector sets $\{\boldsymbol{f}_1, \boldsymbol{f}_2, \ldots, \boldsymbol{f}_k\}$ and $\{\boldsymbol{g}_1, \boldsymbol{g}_2, \ldots, \boldsymbol{g}_k\}$ is so natural that there can be little objection to keeping the name Gram–Schmidt; the context determines immediately whether one or two sets of vectors are involved. Denote by $F^*$ the conjugate transpose of $F$.

**Theorem 1** *Let $F$ and $G$ be complex $n \times k$ matrices, $n \geq k$, such that $G^*F$ permits triangular factorization:*

$$G^*F = \tilde{L}\tilde{D}\tilde{R},$$

*where $\tilde{L}$ and $\tilde{R}$ are unit triangular (left and right), respectively, and $\tilde{D}$ is diagonal. Then there there exist unique $n \times k$ matrices $\tilde{Q}$ and $\tilde{P}$ such that*

$$F = \tilde{Q}\tilde{R}, \quad G = \tilde{P}\tilde{L}^*, \quad \tilde{P}^*\tilde{Q} = \tilde{D}.$$

*Remark.* When $G = F$ the traditional QR factorization is recovered with an unconventional normalization: generally $Q = \tilde{Q}\tilde{D}^{-1/2}$. $F^*F$ permits triangular factorization when, and only when, the leading $k-1$ columns of $F$ are linearly independent.

*Remark.* In practice, when $n = k$ and $\tilde{D}$ is invertible one can omit $Q$ and write $F = \tilde{P}^{-1}(\tilde{D}\tilde{R})$, $G = \tilde{P}\tilde{L}^*$ and still call it *the* Gram–Schmidt factorization. The important feature is the uniqueness of $\tilde{Q}$ and $\tilde{P}$. The columns of $\tilde{Q}$ and rows of $\tilde{P}^*$ form a pair of dual bases for the space of $n$-vectors (columns) and its dual (the row $n$-vectors). There is no notion of orthogonality or inner product here; $p_i^* q_j = 0$ says that $p_i^*$ annihilates $q_j$, $i \neq j$.

We omit the proof of the theorem to save space.

The Gram–Schmidt factorization leads directly to the differential qd algorithms. Let us show how.

**Corollary 1** *Let bidiagonal matrices $L$ and $U$ be given as in Section 2. If $UL$ permits factorization*

$$UL = \hat{L}\hat{D}\hat{R} = \hat{L}\hat{U},$$

*where $\hat{L}$ and $\hat{R}$ are unit bidiagonal, then there exist unique matrices $\tilde{P}$ and $\tilde{Q}$ such that*

$$U = \hat{L}\tilde{P}^*, \quad L = \tilde{Q}\hat{R}, \quad \tilde{P}^*\tilde{Q} = \hat{D}.$$

*Remark.* In words, apply Gram–Schmidt to the columns of $L$ and the rows of $U$, in the natural order, to obtain $\hat{U}$ and $\hat{L}$. Then note that $\hat{U} = \hat{D}\hat{R}$.

Note that if $u_i = 0$, $i < n$, then $UL$ does not permit triangular factorization. However, the theorem allows $u_n = 0$. When $u_n \neq 0$ then $U$ is invertible and so is $\hat{D}$. In this case we can rewrite the factorization as

$$KL = \hat{D}\hat{R} = \hat{U}, \quad UK^{-1} = \hat{L}, \quad K = \hat{D}\tilde{Q}^{-1}.$$

The matrix $K$ is hidden when we just write $UL = \hat{L}\hat{U}$. However, the identification of $K$ with the Gram–Schmidt process goes only half way in the derivation of the dqd algorithm. The nature of the Gram–Schmidt process shows that $\tilde{P}$ and $\tilde{Q}$ are upper Hessenberg matrices. Fortunately $\tilde{Q}$ and $\tilde{P}$ are *special* Hessenberg matrices that depend on only $2n$ parameters, not $n(n-1)/2$. We are going to show that they may be written as the product of $(n-1)$ simple matrices that are non-orthogonal analogues of plane rotations. That means that $L$ may be changed into $\hat{U}$ and $U$ into $\hat{L}$ by a sequence of simple transformations and neither $K$, $\tilde{Q}$ nor $\tilde{P}$ need appear explicitly.

**Definition 1** *A plane transformer, in plane $(i,j)$, $i \neq j$, is an identity matrix except for the entries $(i,i)$, $(i,j)$, $(j,i)$ and $(j,j)$. The $2 \times 2$ submatrix they define must be invertible.*

Let us describe the first minor step in mapping $L \to \hat{U}$, $U \to \hat{L}$. We seek an invertible $2 \times 2$ matrix such that

$$\begin{pmatrix} x & z \\ -y & w \end{pmatrix} \begin{pmatrix} 1 & 0 \\ l_1 & 1 \end{pmatrix} = \begin{pmatrix} \hat{u}_1 & 1 \\ 0 & 1 \end{pmatrix},$$

$$\begin{pmatrix} u_1 & 1 \\ 0 & u_2 \end{pmatrix} \begin{pmatrix} w & -z \\ y & x \end{pmatrix} = \begin{pmatrix} 1 & 0 \\ \hat{l}_1 & * \end{pmatrix} \cdot det,$$

where $det = xw + yz$ and $*$ may be anything. A glance at the last column of the top equation shows that $z = w = 1$. The 0 in the $(2,1)$ entry on the right shows that $y = l_1$ and the 0 in the $(1,2)$ entry in the second equation shows that $x = u_1$. Thus $det = wx + yz = u_1 + l_1 \equiv \hat{u}_1$. From the $(2,1)$ entry of the second equation we learn that

$$u_2 l_1 = u_2 y = \hat{l}_1 det = \hat{l}_1 \hat{u}_1.$$

Finally, and of most interest,

$$* = u_2 x / det = u_2 u_1 / \hat{u}_1.$$

This is the intermediate quantity $d_2$ in dqd and we see it here as something that gets carried to the next minor step. If we write $d_1 = u_1$ we obtain the start of the inner loop of dqd:

$$\begin{aligned} \hat{u}_1 &= d_1 + l_1, \\ \hat{l}_1 &= l_1(u_2/\hat{u}_1), \\ d_2 &= d_1(u_2/\hat{u}_1). \end{aligned}$$

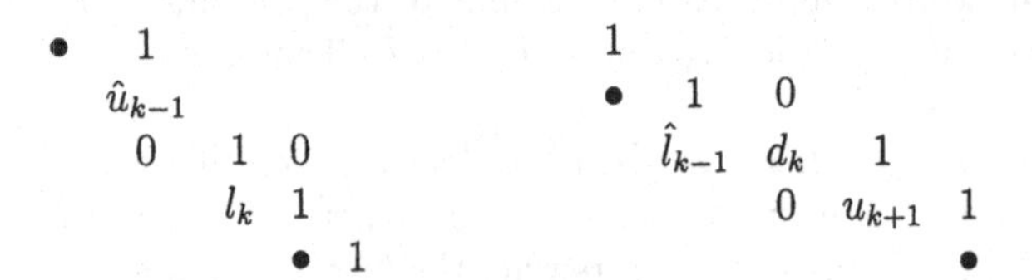

**Figure 6.1** Active entries.

The typical minor step is similar. It is instructive to look at the matrices part way through the transformation $L \to \hat{U}$, $U \to \hat{L}$ as shown in Figure 6.1.

At minor step $k$ the plane transformed is $(k, k+1)$ and the active part of the plane transformer is $\begin{pmatrix} d_k & 1 \\ -l_k & 1 \end{pmatrix}$ on the left and $\begin{pmatrix} 1 & -1 \\ l_k & d_k \end{pmatrix}$ on the right, with $det = \hat{u}_{k+1}$. Finally at the end of minor step $(n-1)$ the trailing $2 \times 2$ submatrices are

$$\begin{pmatrix} \hat{u}_{n-1} & 1 \\ 0 & 1 \end{pmatrix} \quad \text{and} \quad \begin{pmatrix} 1 & 0 \\ \hat{l}_{n-1} & d_n \end{pmatrix} = \begin{pmatrix} 1 & 0 \\ \hat{l}_{n-1} & 1 \end{pmatrix} \begin{pmatrix} 1 & 0 \\ 0 & d_n \end{pmatrix}.$$

If $d_n \neq 0$ we simply multiply row $n$ on the left by $d_n$ and divide column $n$ on the right by $d_n$ as a final similarity transformation. When $d_n = 0$ the matrices $\hat{L}$ and $\hat{R}$ remain invertible. Thus $\hat{u}_n = d_n \hat{r}_{nn} = 0 \cdot 1 = 0$.

So we have derived the dqd algorithm without reference to qd. Of more significance is the fact that the quantities $d_i$, $i = 1, \ldots, n$, provide useful information about $UL$ that qd does not reveal, and so dqd facilitates the choice of shift.

# 7 The meaning of $d_i$

**Theorem 2** *Consider $L$ and $U$ as described in Section 2. If $U$ is invertible then the quantities $d_i$, $i = 1, \ldots, n$, generated by the dqd algorithm applied to $L$ and $U$ satisfy*

$$d_i^{-1} = [(UL)^{-1}]_{ii}, \quad i = 1, \ldots, n.$$

**Proof** The algorithm may be considered as transforming $L$ to $\hat{U}$ by premultiplications and $U$ to $\hat{L}$ by inverse multiplications on the right as described in the previous section. At the end of the $(k-1)$th plane transformation the situation is as indicated below:

$$G_{k-1}L = \begin{bmatrix} \bullet & \bullet & & & \\ & \hat{u}_{k-1} & 1 & & \\ & 0 & 1 & 0 & \\ & & l_k & 1 & \\ & & & \bullet & \bullet \end{bmatrix}, \quad \begin{bmatrix} \bullet & & & & \\ \bullet & 1 & 0 & & \\ & \hat{l}_{k-1} & d_k & 1 & \\ & & 0 & u_{k+1} & 1 \\ & & & \bullet & \bullet \end{bmatrix} = UG_{k-1}^{-1},$$

where

$$G_{k-1} = \Phi_{k-1}\Phi_{k-2}\cdots\Phi_1, \quad \Phi_i \text{ transforms plane } (i, i+1), \quad G_0 = I_n.$$

The striking fact is that row $k$ of $G_{k-1}L$ and column $k$ of $UG_{k-1}^{-1}$ are singletons. If $\boldsymbol{e}_j$ denotes column $j$ of $I_n$ then

$$\boldsymbol{e}_k^t G_{k-1} L = \boldsymbol{e}_k^t, \qquad \boldsymbol{e}_k d_k = U G_{k-1}^{-1} \boldsymbol{e}_k, \quad 1 \le k \le n.$$

Rearranging these equations yields, for $k = 1, 2, \ldots, n$,

$$\begin{aligned} d_k^{-1} &= (\boldsymbol{e}_k^t G_{k-1})(G_{k-1}^{-1}\boldsymbol{e}_k d_k^{-1}) \\ &= (\boldsymbol{e}_k^t L^{-1})(U^{-1}\boldsymbol{e}_k) = [(UL)^{-1}]_{kk}. \end{aligned}$$

□

In the positive case ($l_i > 0$, $u_i > 0$), $UL$ is diagonally similar to a symmetric positive definite matrix.

**Corollary 2** *In the positive case,*

$$\left(\sum_{i=1}^{n} d_i^{-1}\right)^{-1} < \lambda_{\min}(UL) \le \min_i d_i.$$

**Proof** For any matrix $M$ that is diagonally similar to a positive definite symmetric matrix

$$\max_j m_{jj} \le \lambda_{\max}[M] < \operatorname{trace}[M].$$

Take $M = (UL)^{-1}$. □

Even in the general case, as $u_n \to 0$, $\min_i |d_i|$ becomes an increasingly accurate approximation to $|\lambda_{\min}|$.

# 8 Incorporation of shifts

The algorithms and theorems presented so far serve only as background. LR, QR and qd algorithms are only as good as their shift strategies. In practice one uses qds and dqds, the shifted versions of qd and dqd.

The derivation of dqds($\sigma$) in terms of a Gram–Schmidt process is not obvious. Formally we write $UL - \sigma I = (U - \sigma L^{-1})L = \hat{L}\hat{U}$ and apply the Gram–Schmidt process to the columns of $L$ and the rows of $U - \sigma L^{-1}$ to obtain

$$L = G\hat{R}, \quad U - \sigma L^{-1} = \hat{L}F, \quad FG = \hat{D}.$$

Eliminating $G$ yields

$$L = (G\hat{D}^{-1})(\hat{D}\hat{R}) = F^{-1}\hat{U}, \quad U - \sigma L^{-1} = \hat{L}F.$$

At first sight the new term $-\sigma L^{-1}$ appears to spoil the derivation of $F$ as a product of plane transformers. However, it is not necessary to know all the terms of $L^{-1}$ but only the $(i+1, i)$ entries immediately below the main diagonal. The change from the unshifted case is small. The active parts of the two transformations are given by

$$\begin{pmatrix} d_i & 1 \\ -l_i & 1 \end{pmatrix}\begin{pmatrix} 1 & 0 \\ l_i & 1 \end{pmatrix} = \begin{pmatrix} \hat{u}_i & 1 \\ 0 & 1 \end{pmatrix}, \quad \text{as before,}$$

and the new relation

$$\begin{pmatrix} d_i & 1 \\ \sigma l_i & u_{i+1}-\sigma \end{pmatrix}\begin{pmatrix} 1 & -1 \\ l_i & d_i \end{pmatrix} = \begin{pmatrix} 1 & 0 \\ \hat{l}_i & d_{i+1} \end{pmatrix} \cdot det.$$

The last row yields

$$det = d_i + l_i = \hat{u}_i, \quad \text{as before,}$$

$$\begin{aligned} \hat{l}_i \cdot det &= \hat{l}_i\hat{u}_i = \sigma l_i + (u_{i+1}-\sigma)l_i = u_{i+1}l_i, \quad \text{as before,} \\ d_{i+1} \cdot det &= -\sigma l_i + (u_{i+1}-\sigma)d_i = u_{i+1}d_i - \sigma\hat{u}_i. \end{aligned}$$

This is dqds($\sigma$).

If one looks at the two matrices part way through the transformations $L \to \hat{U}$, $U - \sigma L^{-1} \to \hat{L}$, the singleton column in the second matrix (from Theorem 7) has disappeared and the relation of $d_i$ to $(UL)^{-1}$ is more complicated.

**Theorem 3** *Consider L and U as described in Section 2. If U is invertible and $UL - \sigma I = \hat{L}\hat{U}$, then the intermediate quantities $d_i$, $i = 1, \ldots, n$, generated by dqds($\sigma$) applied to L and U, satisfy*

$$\frac{1 + \sigma[(\hat{U}UL)^{-1}]_{k,k-1}}{d_k + \sigma} = [(UL)^{-1}]_{k,k}.$$

We omit the proof.

# 9 Accuracy

The differential qd algorithms dqd and dqds are new to the scene of matrix computations. One feature that makes them attractive is that they seem to be more accurate than their rivals. In particular, in the positive case, all eigenvalues can be found to high relative accuracy as long as the shifts preserve positivity.

Let us begin with an extreme example.

*Example 1.* Take $n = 64$, $u_i = 1$, $i = 1, \ldots, 64$, $l_i = 2^{16} = 65536$, $i = 1, \ldots, 63$. Although $\det(LU) = 1$ the smallest eigenvalue is $O(10^{-304})$.

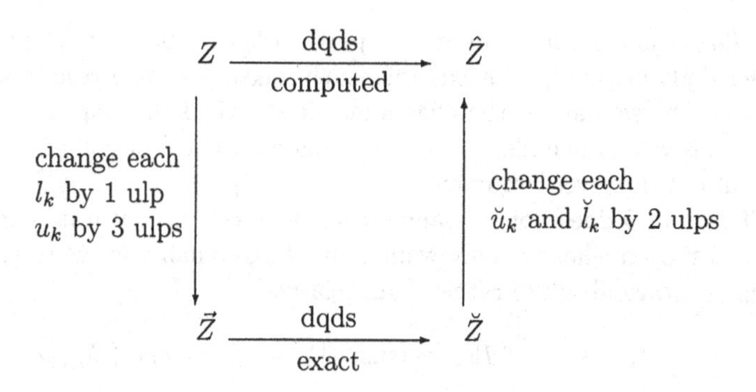


**Figure 9.1** Effects of roundoff.

In just 2 iterations dqd computed $\lambda_{\min}$ to full working precision. In contrast qd returns 0, a satisfactory answer relative to the matrix norm. Yet dqd does preserve the determinant to working accuracy, provided underflow and overflow are absent, while qd, LR and QR do not.

The reason for dqd's accuracy is that $\hat{u}_{64} = d_{64}$ reaches the correct tiny value through 63 multiplications and divisions. There are no subtractions.

A little extra notation is needed to describe the stability results compactly. When there is no need to distinguish $l$'s from $u$'s we follow Rutishauser and speak of a qd array

$$Z = \{u_1, l_1, u_2, l_2, \ldots, l_{n-1}, u_n\}.$$

The right unit for discussing relative errors is the ulp (1 unit in the last place held) since it avoids reference to the magnitudes of the numbers involved. In Example 1 the error in the computed eigenvalue is less than $\frac{1}{2}$ ulp despite $2 \times 63$ divisions and multiplications.

Given $Z$ the dqds algorithm in finite precision produces a representable output $\hat{Z}$. We write this $\hat{Z} = \mathrm{fl}(dqds) \cdot Z$. Now we introduce two ideal qd arrays $\vec{Z}$ and $\breve{Z}$ such that, *in exact arithmetic*, dqds with shift $\sigma$ maps $\vec{Z}$ into $\breve{Z}$. Moreover $\vec{Z}$ is a tiny relative perturbation of $Z$, and $\hat{Z}$ is a tiny relative perturbation of $\breve{Z}$. See Figure 9.1.

The proof of the following result may be found in [2].

**Theorem 4** *In the absence of division by zero, underflow or overflow, the Z diagram commutes and, for all k, $\vec{u}_k$ ($\vec{l}_k$) differs from $u_k$ ($l_k$) by 3 (1) ulps at most, and $\hat{u}_k$ ($\hat{l}_k$) differs from $\breve{u}_k$ ($\breve{l}_k$) by 2 (2) ulps, at most.*

The proof is based on making small changes to $Z$ and $\hat{Z}$ so that the *computed* sequence of $d$'s is exact for $\vec{Z}$ and $\breve{Z}$. There is no requirement of positivity so it is possible to have $\|\hat{Z}\| >>> \|Z\|$. Some people call this a *mixed stability* result because one has to perturb both input and output to get an exact dqds mapping. For example, such mixed accuracy results are the best that can be said about the trigonometric functions in computer systems; the output is within one ulp of the exact trigonometric function of a value within one ulp of the given argument.

Theorem 9 does not guarantee that dqds returns accurate eigenvalues in all cases, even when we only want errors to be small relative to $\|J\|$. We call such accuracy absolute rather than relative:

$$|\lambda_i - \hat{\lambda}_i| < \epsilon\|J\| \quad \text{versus} \quad |\lambda_i - \hat{\lambda}_i| < \epsilon \max\{|\lambda_i|, |\hat{\lambda}_i|\}.$$

In [2] a corollary to Theorem 9 establishes high relative accuracy for all eigenvalues computed by dqds *in the positive case.* The corollary is stated in terms of singular values but the algorithm computes the squares of those singular values, that is, the eigenvalues of a positive definite matrix similar to a product $LU$.

# References

[1] Demmel, J. and W. Kahan (1990) "Accurate singular values of bidiagonal matrices", *SIAM J. Sci. Sta. Comput.* **11**, 873–912.

[2] Fernando, K.V. and B.N. Parlett (1994) "Accurate singular values and differential qd algorithms", *Numerische Mathematik* **67**, 191–229.

[3] Henrici, P. (1958) "The quotient-difference algorithm", *Nat. Bur. Standards Appl. Math. Series* **19**, 23–46.

[4] Parlett, B.N. (1992) "Reduction to tridiagonal form and minimal realizations", *SIAM J. Matrix Anal. and Appls* **13**, 567–593.

[5] Parlett, B.N. (1995) "The New qd Algorithms", *Acta Numerica* **4**, 459–491.

[6] Rutishauser, H. (1954) "Der Quotienten-Differenzen-Algorithmus", *Z. Angew. Math. Phys.* **5**, 233–251.

[7] Rutishauser, H. (1958) "Solution of eigenvalue problems with the LR-transformation", *Nat. Bur. Standards Appl. Math. Series* **49**, 47–81.

[8] Rutishauser, H. (1960) "Über eine kubisch konvergente Variante der LR-Transformation", *Z. Angew. Math. Mech.* **11**, 49–54.

[9] Rutishauser, H. (1990) *Lectures on Numerical Mathematics*, Birkhäuser (Boston).

# On Adjusting Parameters in Homotopy Methods for Linear Programming

*Michael J. Todd*

School of Operations Research and Industrial Engineering, Rhodes Hall, Cornell University, Ithaca, NY 14853, USA[1]

**Abstract**

Several algorithms in optimization can be viewed as following a solution as a parameter or set of parameters is adjusted to a desired value. Examples include homotopy methods in complementarity problems and path-following (infeasible-) interior-point methods. If we have a metric in solution space that corresponds to the complexity of moving from one solution point to another, there is an induced metric in parameter space, which can be used to guide parameter-adjustment schemes. We investigate this viewpoint for feasible- and infeasible-interior-point methods for linear programming.

## 1 Introduction

This paper is concerned with developing guidelines for optimal adjustment of parameters in homotopy or path-following algorithms in optimization, concentrating on interior-point methods for linear programming. The general idea of such algorithms is that, given a particular problem, we regard part of the data, and possibly some additional parameters, as a parameter vector. To obtain a solution corresponding to some fixed value of the parameters, we trace a path as the parameter vector is adjusted from an initial artificial value, for which the corresponding solution is known, to the desired value.
The thesis of our work is that for some such algorithms, in particular for several interior-point methods for linear programming, there is a natural (Riemannian) metric on parameter space with two properties: firstly, the distance between two parameter vectors measures the complexity of obtaining the solution corresponding to the second given that corresponding to the first; and secondly, this complexity can be attained by following the shortest path joining the two vectors in this metric.

[1]Research supported in part by NSF through grant DMS-9505155 and ONR through grant N00014-96-1-0050.

While our focus here is on algorithms for linear programming, Mike Powell's work in optimization has been mainly concerned with nonlinear programming. However, the methods we consider use primarily ideas from nonlinear programming, particularly those of the classical barrier function approach to constrained optimization. In addition, Powell himself has made two very fine contributions to interior-point algorithms for linear programming: an analysis of Karmarkar's original algorithm [7] for discretizations of a simple two-variable semi-infinite programming problem [20, 19], and a general convergence proof [21] for Polyak's modified barrier method [18]. In the first, the concern was complexity: Powell showed that the number of iterations required could be close to the bound established by Karmarkar; in the second, the topic was proving convergence of a (primal) method that did not require an initial strictly feasible solution. Our interest here also revolves around these two themes of complexity and infeasible-interior-point methods, but now from a primal-dual viewpoint.

There is another connection to Powell's work, indeed to his pioneering paper with Fletcher [4] on Davidon's variable metric algorithm for unconstrained minimization [2]. These papers introduced the famous DFP update formula and more generally gave birth to the field of quasi-Newton methods. The idea of a variable metric is key in this and also in our work. For unconstrained minimization, the Hessian of the objective function or a positive definite approximation to it defines a local norm at the current solution, and steepest descent with respect to this metric permits good progress in decreasing the function. As the iterates move, the curvature of the function changes, and the appropriate metric to yield good behavior also changes.

In interior-point methods a variable metric is also fundamental, but here it is used to describe the *constraint set* $K$ rather than the objective function. This is different from incorporating curvature from the constraint functions into the objective function via the Lagrangian as in sequential quadratic programming, for example, because curvature is induced even by linear constraints; the metric reflects the geometry of *all* the constraints, not just those estimated to be active at the solution. Just as in general relativity the presence of all (but especially nearby) masses curves space-time, so here the presence of all (but especially nearby) constraint boundaries determines the local geometry.

To give an example, for the nonnegative orthant $K := \mathbb{R}^n_+$, at every point $x$ of the positive orthant $\operatorname{int} K = \mathbb{R}^n_{++}$ we can define a local norm by $\|v\|_x := \|X^{-1}v\|_2$, where here and below $X$ denotes the diagonal matrix containing the components of $x$. First, this norm reflects in some sense the shape of $K$: $\{x + v : \|v\|_x < 1\} \subseteq \operatorname{int} K$. Second, it is the ellipsoidal norm defined by the

Hessian of a barrier function for $K$, namely the logarithmic barrier function

$$\phi(x) := -\sum_{j=1}^{n} \ln(x^{(j)}), \tag{1.1}$$

where $x^{(j)}$ denotes the $j$th component of $x$: $\|v\|_x = (v^\top \phi''(x)v)^{1/2}$. Third, if we are close to a desired point in this norm, we can approximate it well by a single Newton step: if $\bar{x}$ solves $\min\{\mu^{-1}c^\top x + \phi(x) : Ax = b, x \in K\}$, and we have a point $x \in \operatorname{int} K$ with $\|\bar{x} - x\|_x \leq \delta$ for some absolute constant $\delta < 1$, then one Newton step from $x$ will give a very good approximation to $\bar{x}$. (Results of this form have been proved in a very general framework by Nesterov and Nemirovskii [14].)

Actually, we shall use not only local norms, but also the induced Riemannian metrics (the distance between two points is the length of a shortest path, measured using the local norm, connecting them) in what follows. The resulting Riemannian metric in parameter space will have the two properties mentioned at the beginning.

This paper is organized as follows. In Section 2 we describe the homotopy methods of interest, first in an abstract framework and then for linear programming. Then Section 3 discusses metrics in solution space, first in general and then for linear programming, and then shows under a certain key hypothesis how the complexity of a path-following method can be bounded by a constant times the induced distance between two parameter vectors. In Section 4 we calculate the local norm in parameter space for linear programming, and Section 5 then finds some shortest paths. Some of these challenge the conventional wisdom that recommends approaching feasibility faster than (or at the same rate as) optimality in infeasible-interior-point methods. Finally, Section 6 sketches a justification for the crucial hypothesis concerning the metrics in the case of linear programming.

A forthcoming paper will show how the same ideas can be applied to pivoting algorithms and to interior-point methods for more general convex programming problems, and will contain proofs of some of the results, proofs that are omitted here.

Related ideas of how to adjust parameters in interior-point methods appear in the target-following work of Mizuno [11] and Jansen, Roos, Terlaky and Vial [5, 6]. Of course, the key ideas of Nesterov and Nemirovskii [14] on how barrier functions induce metrics with computational significance were also an important catalyst. While this research was in progress, I became aware of related later work by Nesterov and Nemirovskii [15] on moving efficiently in a multi-parameter surface. I would also like to acknowledge very helpful conversations with Jim Renegar on this approach, with Clovis Gonzaga on infeasible-interior-point methods, and with Gongyun Zhao, who has had very similar ideas, on appropriate metrics.

Our notation is mostly standard. We will use upper case letters (like $X$, $S$, and $W$) to denote the diagonal matrices containing the components of the corresponding vectors (like $x$, $s$, and $w$), and $e$ to denote the $n$-vector of ones, so that $Xe = x$, etc. Thus $XSe$ denotes the vector whose components are the products of those of $x$ and $s$, but we shall also use $xs$ to denote the same vector, and similarly $x^{1/2}$, $x^{-1}$, and $\ln x$ to denote the vectors of square roots, reciprocals, and logarithms of the components of $x$, etc. Sequences will be indicated by subscripts, and components by superscripts inside parentheses, as in (1.1) above.

# 2 Homotopy methods for mathematical programming

Here we describe formally the class of methods we are dealing with and show how interior-point algorithms for linear programming are included.

Solution vectors $z$ lie in some subset $\mathcal{Z}$ of a Euclidean space, while parameter vectors $\lambda$ lie in a subset $\Lambda$ of another Euclidean space. (Perhaps it would be more accurate to say subsets of finite-dimensional real vector spaces, because we shall use metrics (Riemannian metrics) on these spaces, which may be very different from Euclidean metrics, in order to measure the complexity of homotopy algorithms.) The solution and the parameter are related implicitly by the equation

$$F(z, \lambda) = 0. \tag{2.1}$$

We assume that $F$ is continuously differentiable and that for each $\lambda \in \Lambda$ there is a unique $z = z(\lambda) \in \mathcal{Z}$ satisfying (2.1). We also assume that the partial derivative with respect to $z$, $F_z(z(\lambda), \lambda)$, is nonsingular for any $\lambda \in \Lambda$.

We suppose we know $z_0 := z(\lambda_0)$ (or a good approximation to it) for some $\lambda_0 \in \Lambda$, and we seek $z_1 := z(\lambda_1)$ for some $\lambda_1 \in \Lambda$. The basic idea of a homotopy or path-following method is to define a path $\gamma : [0,1] \to \Lambda$ from $\lambda_0$ to $\lambda_1$ (i.e., with $\gamma(i) = \lambda_i$ for $i = 0, 1$), and to trace or approximate the solutions $\zeta(t) := z(\gamma(t))$ as $t$ goes from 0 to 1. The basic mechanics of the method determine how the tracing is to be performed (usually via Newton or Newton-like steps), whereas the key high-level question is how the path $\gamma$ is to be chosen. Our criterion throughout will be the *complexity* of obtaining $z_1$, i.e., of tracing the path $\zeta$. Simple examples show that the linear interpolation

$$\gamma(t) := (1-t)\lambda_0 + t\lambda_1 \tag{2.2}$$

may not be best according to this criterion.

While the focus here is on short-step path-following methods, we suspect that the paths selected will also be appropriate for adaptive- or long-step methods, where the parameter is moved at each iteration as far along the path as a single Newton step can track accurately or where a long step along the path is attempted by taking a sequence of Newton steps.

We remark that a parallel theory can be developed for the case that $F(z, \lambda) = f(z) - \lambda$, with $f$ piecewise linear. In this case, pivots replace Newton steps at each iteration, and the metrics to be defined are not Riemannian but basically count the number of pieces of linearity encountered. Lemke's method [10] can be viewed as a homotopy method in this framework. For simplicity, we confine ourselves here to the continuously differentiable case. This still encompasses a great variety of computational algorithms, for example smooth continuation methods for zero-finding or fixed-point problems. Our motivation comes from interior-point, and in particular infeasible-interior-point methods, for linear programming and its extensions, originating with Karmarkar [7]. Here, due largely to the pioneering work of Nesterov and Nemirovskii [14], a great deal is known about the behavior of a Newton step at each iteration based on a fundamental metric defined by a barrier function. Our aim is to formalize the use of these metrics and use them to suggest 'good' paths $\gamma$ to be used in path-following methods.

Let us therefore describe the framework of path-following interior-point methods for linear programming. In fact, our development also applies to similar methods for more general conic problems as studied by Nesterov and Nemirovskii [14], where the nonnegative orthant is replaced by a closed convex solid pointed cone. If the cone has a *self-scaled* barrier in the sense of Nesterov and Todd [16, 17], then similar results can be established. Such problems include those of semidefinite programming, which have been much studied recently.

## 2.1 Interior-point methods for linear programming

Suppose we wish to find an optimal solution to the problem

$$
(\mathrm{LP}) \qquad \begin{array}{rl} \min_x & c_1^\top x \\ & Ax = b_1, \\ & x \geq 0, \end{array}
$$

and its dual

$$
(\mathrm{LD}) \qquad \begin{array}{rl} \max_{y,s} & b_1^\top y \\ & A^\top y + s = c_1, \\ & s \geq 0, \end{array}
$$

where $A \in \mathbb{R}^{m\times n}$, $b_1 \in \mathbb{R}^m$, $c_1 \in \mathbb{R}^n$, $x, s \in \mathbb{R}^n$, and $y \in \mathbb{R}^m$. We always suppose $A$ fixed and, without loss of generality, of full row rank. The right-hand-side vector and the cost vector may be fixed, but often will be part of the parameter vector. Our approach is via solutions to the system

$$\begin{array}{rcrcrcl} Ax & & & - & b & = & 0, \quad (x \geq 0) \\ A^\top y & + & s & - & c & = & 0, \quad (s \geq 0) \\ & & XSe & - & v^2 & = & 0 \end{array} \tag{2.3}$$

for a sequence of values of $b, c$, and $v$. Recall that our notation here is that $e$ always denotes a vector of ones in $\mathbb{R}^n$, while $X$ and $S$ are diagonal matrices with $Xe = x$ and $Se = s$; $v^2$ denotes the vector whose components are the squares of those of $v \in \mathbb{R}^n$ (the *target* vector). We also write $xs$ for $XSe$ when no confusion can result.

Note first that the optimality conditions for (LP) and (LD) are exactly (2.3) for $v = 0$. Interior-point methods consider solutions for vectors $v > 0$, in which case $x$ and $s$ are both positive (interior points of the cone $\mathbb{R}^n_+$). We will therefore ignore the nonnegativity conditions in (2.3), and write the left-hand side of these equations as $F(z, \lambda)$, where always $z := (x, y, s) \in \mathcal{Z} := \mathbb{R}^n_{++} \times \mathbb{R}^m \times \mathbb{R}^n_{++}$.

Note that $F_z(z, \lambda)$ is then

$$\begin{bmatrix} 0 & A & 0 \\ A^\top & 0 & I \\ 0 & S & X \end{bmatrix},$$

which is nonsingular for any $x, s > 0$; it is also well-known that there is a unique solution $(x, y, s) \in \mathcal{Z}$ for any $b, c$, and $v > 0$ for which $Ax = b$ and $A^\top y + s = c$ have solutions with $x, s > 0$.

There are several choices for $\Lambda$. Suppose first that (LP) and (LD) have *strictly feasible* solutions, with $x$ and $s$ positive. Then there is a very important path in the set of feasible solutions to (LP) and (LD), called the *central path*. It consists of all solutions to (2.3) with $b = b_1$, $c = c_1$, and $v^2 = \mu e$ for some positive $\mu$. Suppose we have strictly feasible solutions $x_0$ and $(y_0, s_0)$, with $x_0 s_0$ close to $\mu_0 e$ for some $\mu_0 > 0$. Then we can approximate a solution to (LP) and (LD) by following the central path; this corresponds to choosing $\lambda := \mu \in \Lambda := (0, \infty)$, fixing $b$ and $c$ as $b_1$ and $c_1$, and replacing $v^2$ by $\mu e$. In this case our parameter is one-dimensional, so the question of choosing the path $\gamma$ is moot, but we will still be able to say something about the complexity of moving from $\mu = \mu_0$ to $\mu = \mu_1$ in terms of the length (suitably defined) of this path.

Recall that for any feasible solutions $x$ and $(y, s)$ to (LP) and (LD),

$$c^\top x - b^\top y = (A^\top y + s)^\top x - (Ax)^\top y = s^\top x \geq 0,$$

so the primal objective value is always at least the dual objective value. The difference is called the duality gap, and is zero at optimality. If $z = (x, y, s)$ solves (2.3) with $v^2 = \mu e$, then the duality gap is

$$s^\top x = e^\top XSe = e^\top(\mu e) = n\mu, \tag{2.4}$$

so finding a solution for a small value $\mu_1$ guarantees a small duality gap. Hence central path-following algorithms (e.g., Monteiro–Adler [13] and Kojima–Mizuno–Yoshise [8, 9]) are included in our framework.

Next suppose we again have strictly feasible solutions $x_0$ and $(y_0, s_0)$, but $x_0 s_0$ is not close to a multiple of $e$. We could perform some 'centering' steps to approach the central path and then proceed as above, but it may be more efficient to use a target-following method. Thus we again fix $b$ and $c$ as $b_1$ and $c_1$, but now use $\lambda := v \in \Lambda := \mathbb{R}^n_{++}$ as our parameter vector. Note that the duality gap for the corresponding solution is now

$$s^\top x = e^\top XSe = e^\top v^2 = \|v\|_2^2, \tag{2.5}$$

so we want to adjust $\lambda = v$ from its initial value $v_0$ with $x_0 s_0 = v_0^2$ to some $v_1$ of small norm. Thus the methods of Mizuno [11] and Jansen–Roos–Terlaky–Vial [5, 6] are embraced.

If we do not have strictly feasible solutions, we can choose an arbitrary starting point $z_0 = (x_0, y_0, s_0) \in \mathcal{Z}$ and set $b_0 = Ax_0$, $c_0 = A^\top y_0 + s_0$. There remain several choices for $\lambda$. If $x_0 s_0$ is close to $\mu_0 e$ for some $\mu_0 > 0$, we can choose $\lambda := (b, c, \mu) \in \Lambda := A(\mathbb{R}^n_{++}) \times (A^\top(\mathbb{R}^m) + \mathbb{R}^n_{++}) \times \mathbb{R}_{++}$, and follow solutions to (2.3), with $v^2$ replaced by $\mu e$, as $\lambda$ moves from $\lambda_0 := (b_0, c_0, \mu_0)$ to $\lambda_1 := (b_1, c_1, \mu_1)$ for some suitably small $\mu_1$. If not, we can choose $\lambda := (b, c, v)$ in the obvious $\Lambda$ and proceed similarly. We can also replace $b$ and $c$ by $(1-\theta)b_0 + \theta b_1$ and $(1-\theta)c_0 + \theta c_1$, and use $\lambda := (\theta, \mu)$ or $\lambda := (\theta, v)$ as our parameter. Hence we include most *infeasible-interior-point* methods; see, e.g., Mizuno–Todd–Ye [12] and the references therein.

# 3 Metrics reflecting complexity

In this section we show how, given a metric $d_{\mathcal{Z}}$ on $\mathcal{Z}$ indicating the complexity of moving the solution vector, we can infer a metric $d_\Lambda$ on $\Lambda$ so that we can obtain (a good approximation to) $z(\lambda_1)$ from (a good approximation to) $z(\lambda_0)$ in $O(d_\Lambda(\lambda_0, \lambda_1))$ basic steps, and the corresponding path $\gamma$ in $\Lambda$ is the shortest path in $\Lambda$ according to $d_\Lambda$.

We will suppose that $\mathcal{Z}$ is a differentiable manifold endowed with a Riemannian structure, whose distance reflects the complexity of moving between two different solution vectors. We then pull this metric back to $\Lambda$, which is also assumed to be a differentiable manifold. Then the distance in $\Lambda$ from $\lambda_0$ to $\lambda_1$ will represent the 'best' complexity of a path-following method to approximate $z(\lambda_1)$ given $z(\lambda_0)$, and the corresponding shortest path a desired path to follow in $\Lambda$ to achieve this complexity. We illustrate the development using linear programming.

Let $\mathcal{Z}$ be a differentiable manifold equipped with a Riemannian structure (see, for example, Boothby [1]), so that for each $z \in \mathcal{Z}$, there is an inner product $\langle\cdot,\cdot\rangle_z$ defined on the tangent space at $z$. Thus for every tangent vector $\dot{z}$ at $z$ (we use $\dot{z}$ instead of $dz$ to simplify the notation; it does not necessarily indicate the derivative of a path, but no confusion should result), we can define its $z$-norm by

$$\|\dot{z}\|_z := \langle\dot{z},\dot{z}\rangle_z^{1/2}. \tag{3.1}$$

We call this the *local norm at $z$*. Then given a piecewise smooth path $\zeta : [0,1] \to \mathcal{Z}$, we define its length by

$$\ell_{\mathcal{Z}}(\zeta) := \int_0^1 \|\dot{\zeta}(t)\|_{\zeta(t)} dt. \tag{3.2}$$

Finally, the distance between $z_0$ and $z_1$ in $\mathcal{Z}$ is defined to be

$$d_{\mathcal{Z}}(z_0, z_1) := \inf_{\zeta} \ell(\zeta), \tag{3.3}$$

where the infimum is taken over all piecewise smooth paths $\zeta$ from $z_0$ to $z_1$, i.e., with $\zeta(i) = z_i$, $i = 0, 1$. We also define a modified metric

$$d'_{\mathcal{Z}}(z_0, z_1) := d_{z(\Lambda)}(z_0, z_1) := \inf_{\zeta} \ell(\zeta)$$

for points $z_0 = z(\lambda_0)$ and $z_1 = z(\lambda_1)$, where now the infimum is taken over piecewise smooth paths $\zeta : [0,1] \to z(\Lambda)$ from $z_0$ to $z_1$.

Given a point $\bar{z} = z(\lambda) \in \mathcal{Z}$, we call $z \in \mathcal{Z}$ an *$\eta$-approximation* to $\bar{z}$ if either

$$d_{\mathcal{Z}}(z, \bar{z}) \leq \eta \quad \text{or} \quad \|z - \bar{z}\|_{\bar{z}} \leq \eta. \tag{3.4}$$

The second choice is often more convenient, and we shall use it in our application to linear programming, but it requires that $z - \bar{z}$ be a tangent vector at $\bar{z}$; this certainly holds if $\mathcal{Z}$ is an open set (as in linear programming), so that the tangent space at any point is the Euclidean space in which $\mathcal{Z}$ is embedded. Note that given only $z$ and $\lambda$, it may be hard to recognize whether (3.4) holds; we will comment further on this in Section 6.

The key requirement we make on our metric is:

**Hypothesis 1** *There are constants $\epsilon > 0$ and $\eta \geq 0$ with the following property. Suppose $\bar{z} = z(\lambda)$ and $\bar{z}_+ = z(\lambda_+)$ for some $\lambda, \lambda_+ \in \Lambda$, and suppose $z$ is an $\eta$-approximation to $\bar{z}$. Then, as long as*

$$d'_{\mathcal{Z}}(\bar{z}, \bar{z}_+) \leq \epsilon, \tag{3.5}$$

*the Newton step from $z$ for the system $F(\cdot, \lambda_+) = 0$ is well-defined and yields an $\eta$-approximation $z_+$ to $\bar{z}_+$.*

In other words, if we have a good approximation $z$ to the solution vector corresponding to the parameter vector $\lambda$, and we adjust the parameter vector so that the solution vector moves a small amount ($\epsilon$ in the Riemannian metric), then we can recover a good approximation to the new solution vector by performing a single Newton step.

Verifying that Hypothesis 1 holds in a particular setting can be arduous. For now, we merely define the metric for the cases of linear programming, postponing any discussion of its validity until the final section.

## 3.1 Linear programming

Here $\mathcal{Z} = \mathbb{R}^n_{++} \times \mathbb{R}^m \times \mathbb{R}^n_{++}$, and at any $z \in \mathcal{Z}$ the tangent space is just $\mathbb{R}^n \times \mathbb{R}^m \times \mathbb{R}^n$. First we define primal and dual norms in both primal ($x$) and dual ($s$) spaces. We set

$$\|v\|_x := \|X^{-1}v\|_2, \quad \|u\|_x^* := \|Xu\|_2, \tag{3.6}$$

$$\|u\|_s := \|S^{-1}u\|_2, \quad \|v\|_s^* := \|Sv\|_2. \tag{3.7}$$

Note the simple property that, if $x \in \mathbb{R}^n_{++}$ and $v \in \mathbb{R}^n$ with $\|v\|_x < 1$, then $x \pm v \in \mathbb{R}^n_{++}$, and similarly for $s$. To define the Riemannian structure at $z = (x, y, s) \in \mathcal{Z}$, we first compute $\mu := s^\top x/n$ and $w := x^{1/2}s^{-1/2}$, $t := s^{1/2}x^{-1/2}$, and then for any $\dot{z}_i = (\dot{x}_i, \dot{y}_i, \dot{s}_i)$ in the tangent space, $i = 1, 2$, we set

$$\langle \dot{z}_1, \dot{z}_2 \rangle_z := \mu^{-1}(\dot{x}_1^\top W^{-2}\dot{x}_2 + \dot{s}_1^\top T^{-2}\dot{s}_2), \tag{3.8}$$

so that

$$\|\dot{z}\|_z = \mu^{-1/2}(\|\dot{x}\|_w^2 + \|\dot{s}\|_w^{*2})^{1/2}, \tag{3.9}$$

where $\dot{z} = (\dot{x}, \dot{y}, \dot{s})$ (this notation will be implicit from now on). This is not strictly a norm; it becomes one if we eliminate the free variables $y$, but they are convenient to retain.

To motivate the use of $w$ here, let us note that the norm $\|\cdot\|_x$ and its dual $\|\cdot\|_x^*$ arising from the primal solution $x$ are generally unrelated to the norm $\|\cdot\|_s$ and its dual $\|\cdot\|_s^*$ arising from the dual solution. To effect a compromise, we use the intermediate vector $w$. Let us recall the logarithmic barrier function

$$\phi(x) := -\sum_{j=1}^{n} \ln(x^{(j)}), \tag{3.10}$$

where $x^{(j)}$ denotes the $j$th component of $x$. Then $\|v\|_x = \langle\phi''(x)v, v\rangle^{1/2}$ and $\|u\|_x^* = \langle u, [\phi''(x)]^{-1}u\rangle^{1/2}$, and similarly for the norms associated with $s$, $w$, and $t$. Note that $\phi''(w)x = s$ (and similarly $\phi''(t)s = x$, with $\phi''(t) = [\phi''(w)]^{-1}$), so that the norms defined by $w$ and $t$ are dual and are symmetric between the primal and the dual. In interior-point terminology, methods based on the norm $\|\cdot\|_x$ are called primal-scaling methods, those using $\|\cdot\|_s$ are called dual-scaling, while those using $\|\cdot\|_w$ are called (symmetric) primal-dual-scaling methods.

In one important special case, these norms are all related. If $z$ is *central*, so that $xs = \mu e$, then $w = \mu^{-1/2}x$ and (3.9) simplifies to

$$\|\dot{z}\|_z = (\|\dot{x}\|_x^2 + \|\dot{s}\|_s^2)^{1/2}. \tag{3.11}$$

Then $\|\dot{z}\|_z < 1$ implies that $\|\dot{x}\|_x$ and $\|\dot{s}\|_s$ are less than 1, so that $z \pm \dot{z} \in \mathcal{Z}$. This does not hold in general, but if $xs \geq \theta^2\mu e$ for some $\theta \in (0, 1]$, it is easy to see that $\|\dot{z}\|_z < \theta$ implies that $z \pm \dot{z} \in \mathcal{Z}$.

From this norm, we define the metric as above.

## 3.2 The metric on $\Lambda$

Given a metric on $\mathcal{Z}$, we pull it back to get a metric on $\Lambda$ as follows. For any $\lambda \in \Lambda$ and $\dot{\lambda}$ in the tangent space to $\Lambda$ at $\lambda$, we let $z = z(\lambda)$ and compute $\dot{z}$ from

$$F_z(z, \lambda)\dot{z} + F_\lambda(z, \lambda)\dot{\lambda} = 0 \tag{3.12}$$

(this uniquely defines $\dot{z}$ from our assumptions below (2.1)). Then we set

$$\|\dot{\lambda}\|_\lambda := \|\dot{z}\|_z; \tag{3.13}$$

we could similarly define $\langle\cdot,\cdot\rangle_\lambda$, but it is not needed. In this way we get a Riemannian structure and hence a metric on $\Lambda$.

Let $\gamma : [0, 1] \to \Lambda$ be a smooth path in $\Lambda$, and let $\zeta(t) = z(\gamma(t))$ define the corresponding path in $\mathcal{Z}$. Then the implicit function theorem shows that $\dot{\zeta}(t)$ is the $\dot{z}$ corresponding to $\dot{\lambda} = \dot{\gamma}(t)$, and hence from (3.13) that

$$\|\dot{\gamma}(t)\|_{\gamma(t)} = \|\dot{\zeta}(t)\|_{\zeta(t)}.$$

It follows that

$$\ell_\Lambda(\gamma) = \ell_{\mathcal{Z}}(\zeta), \tag{3.14}$$

and hence

$$d_\Lambda(\lambda_0, \lambda_1) = d'_{\mathcal{Z}}(z(\lambda_0), z(\lambda_1)). \tag{3.15}$$

From this we obtain the result which justifies our interest in this metric.

**Theorem 2** *Given Hypothesis 1, there is a path-following method which, from a good approximation to $z_0 = z(\lambda_0)$, obtains a good approximation to $z_1 = z(\lambda_1)$, and requires*

$$O(d_\Lambda(\lambda_0, \lambda_1)) \tag{3.16}$$

*Newton steps.*

**Proof** Let $\gamma$ be a path in $\Lambda$ from $\lambda_0$ to $\lambda_1$ of length at most $2d_\Lambda(\lambda_0, \lambda_1)$, and let $\zeta(t) := z(\gamma(t))$, so that the length of $\zeta$ is also at most $2d_\Lambda(\lambda_0, \lambda_1)$. Now divide $\zeta$ into intervals $z_0 = \bar{z}_{(0)}, \bar{z}_{(1)}, \ldots, \bar{z}_{(k)} = z_1$, with $d'_Z(\bar{z}_{(i-1)}, \bar{z}_{(i)}) \leq \epsilon$ and $\bar{z}_{(i)} = z(\lambda_{(i)})$ for $1 \leq i \leq k$. Then $k \leq 2d_\Lambda(\lambda_0, \lambda_1)/\epsilon = O(d_\Lambda(\lambda_0, \lambda_1))$. We suppose we have an $\eta$-approximation $z_{(0)}$ to $\bar{z}_{(0)} = z_0$. In general, assuming that we have an $\eta$-approximation $z_{(i-1)}$ to $\bar{z}_{(i-1)}$, we can obtain according to Hypothesis 1 an $\eta$-approximation $z_{(i)}$ to $\bar{z}_{(i)}$ by taking a single Newton step for $F(\cdot, \lambda_{(i)}) = 0$ from $z_{(i-1)}$. Hence, by induction, in $k$ Newton steps we will obtain an $\eta$-approximation $z_{(k)}$ to $\bar{z}_{(k)} = z_1$. □

# 4 The metric on parameter space for linear programming

Let us first consider the general infeasible-interior-point target-following method, so that $\lambda = (b, c, v)$. The corresponding solution is $z = (x, y, s) = z(\lambda)$, at which the metric is defined by (3.9). Let us denote

$$H = W^{-2} = X^{-1}S,$$

so that (3.9) can be written

$$\|\dot{z}\|_z = \mu^{-1/2}(\dot{x}^\top H\dot{x} + \dot{s}^\top H^{-1}\dot{s})^{1/2}. \tag{4.1}$$

Suppose we are given a displacement $\dot{\lambda} = (\dot{b}, \dot{c}, \dot{v})$ in parameter space. We then compute the corresponding displacement $\dot{z}$ via (3.12), which becomes

$$\begin{array}{rcrcl} A\dot{x} & & & = & \dot{b} \\ A^\top\dot{y} & + & \dot{s} & = & \dot{c} \\ S\dot{x} & + & X\dot{s} & = & 2v\dot{v}, \end{array} \tag{4.2}$$

or equivalently

$$\begin{array}{rcrcl} A\dot{x} & & & = & \dot{b} \\ A^\top\dot{y} & + & \dot{s} & = & \dot{c} \\ H\dot{x} & + & \dot{s} & = & 2W^{-1}\dot{v}, \end{array} \tag{4.3}$$

For later use, we find the solution $\dot{z}$ to (4.3) where $2W^{-1}\dot{v}$ is replaced with $\dot{g}$.

Let us define

$$J := (AH^{-1}A^\top)^{-1}, \quad Q := A^\top JA, \quad P := H^{-1} - H^{-1}QH^{-1}. \tag{4.4}$$

Then it is easy to check that

$$\begin{array}{l} J, P \text{ and } Q \text{ are symmetric;} \\ AP = 0, \quad AH^{-1}Q = A; \\ PQ = 0, \quad QP = 0; \\ HP + QH^{-1} = H^{1/2}PH^{1/2} + H^{-1/2}QH^{-1/2} = I; \\ PHP = P, \quad QH^{-1}Q = Q. \end{array} \tag{4.5}$$

From these properties, it is straightforward to confirm that the solution to (4.3) (with its third right-hand side replaced by $\dot{g}$) is

$$\begin{aligned} \dot{x} &= P(\dot{g} - \dot{c}) + H^{-1}A^\top J\dot{b}, \\ \dot{y} &= -JAH^{-1}(\dot{g} - \dot{c}) + J\dot{b}, \\ \dot{s} &= HP\dot{c} + QH^{-1}\dot{g} - A^\top J\dot{b}. \end{aligned} \tag{4.6}$$

From this we calculate, using (4.5) again,

$$\begin{aligned} \dot{x}^\top H\dot{x} + \dot{s}^\top H^{-1}\dot{s} = {} & (\dot{g} - \dot{c})^\top P(\dot{g} - \dot{c}) + \dot{b}^\top J\dot{b} \\ & + \dot{c}^\top P\dot{c} + \dot{g}^\top H^{-1}QH^{-1}\dot{g} + \dot{b}^\top J\dot{b} \\ & - 2\dot{b}^\top JAH^{-1}\dot{g}. \end{aligned} \tag{4.7}$$

We simplify this expression using the new variable

$$\dot{p} := \dot{c} - QH^{-1}\dot{c} + A^\top J\dot{b}. \tag{4.8}$$

Note that $AH^{-1}\dot{p} = \dot{b}$. Moreover, if we replace $\dot{c}$ in (4.3) by $\dot{p}$ (which differs by a vector in the range of $A^\top$), the only change in the solution occurs in $\dot{y}$. Thus the quantity in (4.7) remains unchanged. Let us therefore replace $\dot{b}$ by $AH^{-1}\dot{p}$ and $\dot{c}$ by $\dot{p}$ in (4.7), to get

$$\begin{aligned} \dot{x}^\top H\dot{x} + \dot{s}^\top H^{-1}\dot{s} &= (\dot{g} - \dot{p})^\top P(\dot{g} - \dot{p}) + \dot{p}^\top H^{-1}QH^{-1}\dot{p} \\ &\quad + \dot{p}^\top P\dot{p} + \dot{g}^\top H^{-1}QH^{-1}\dot{g} \\ &\quad + \dot{p}^\top H^{-1}QH^{-1}\dot{p} - 2\dot{p}^\top H^{-1}QH^{-1}\dot{g} \\ &= (\dot{g} - \dot{p})^\top (P + H^{-1}QH^{-1})(\dot{g} - \dot{p}) + \dot{p}^\top (P + H^{-1}QH^{-1})\dot{p} \\ &= (\dot{g} - \dot{p})^\top H^{-1}(\dot{g} - \dot{p}) + \dot{p}^\top H^{-1}\dot{p}. \end{aligned}$$

We conclude that

$$\begin{aligned} \|\dot{z}\|_z &= \mu^{-1/2}(\dot{x}^\top H\dot{x} + \dot{s}^\top H^{-1}\dot{s})^{1/2} \\ &= \mu^{-1/2}((\dot{g} - \dot{p})^\top H^{-1}(\dot{g} - \dot{p}) + \dot{p}^\top H^{-1}\dot{p})^{1/2}. \end{aligned} \tag{4.9}$$

The expression (4.9) is useful in the general case of conic programming, but for linear programming it can be further simplified. Note that $\mu := s^\top x/n = e^\top XSe/n = \|v\|_2^2/n$. Also, $H^{-1/2}\dot{g} = 2\dot{v}$. Thus we obtain

$$\|\dot{z}\|_z = \frac{n^{1/2}}{\|v\|_2}(\|2\dot{v} - \dot{q}\|_2^2 + \|\dot{q}\|_2^2)^{1/2}, \tag{4.10}$$

where $\dot{q} := H^{-1/2}\dot{p}$.

We have therefore proved:

**Theorem 3** *For the case of linear programming with parameter space* $\Lambda = \{(b, c, v)\}$, *the local norm (3.13) defining the metric is given by*

$$\|(\dot{b}, \dot{c}, \dot{v})\|_{(b,c,v)} = \frac{n^{1/2}}{\|v\|_2}(\|2\dot{v} - \dot{q}\|_2^2 + \|\dot{q}\|_2^2)^{1/2}, \tag{4.11}$$

*where* $z = (x, y, s) := z(b, c, v)$ *and*

$$\dot{q} := (I - H^{-1/2}A^\top(AH^{-1}A^\top)^{-1}AH^{-1/2})H^{-1/2}\dot{c} + H^{-1/2}A^\top(AH^{-1}A^\top)^{-1}\dot{b},$$

*with* $H := X^{-1}S$.

Let us consider various special cases.

First we suppose we have feasible solutions so that we are addressing feasible-interior-point target-following methods. Then $b$ and $c$ remain fixed as $b_1$ and $c_1$, and the natural parameter space is $\Lambda = \{v\} := \mathbb{R}^n_{++}$. We can identify this with $\{(b_1, c_1, v)\} = \{b_1\} \times \{c_1\} \times \mathbb{R}^n_{++}$, a subset of the parameter space considered above. To obtain the corresponding metric, we only need to set $\dot{b}$ and $\dot{c}$ to zero in (4.11). Thus $\dot{q}$ is zero and we find

$$\|\dot{v}\|_v = 2n^{1/2}\frac{\|\dot{v}\|_2}{\|v\|_2}. \tag{4.12}$$

We will determine shortest-path geodesics corresponding to the metric corresponding to (4.12) in the next section.

Second, let us return to the infeasible case but suppose that we are concerned with infeasible-interior-point central-path-following methods. Then the natural parameter space is $\Lambda := \{(b, c, \mu)\}$ but we can identify this with the subset $\{(b, c, \mu^{1/2}e)\}$ of our space $\{(b, c, v)\}$ above. Corresponding to a displacement $\dot{\lambda} = (\dot{b}, \dot{c}, \dot{\mu})$ in $\Lambda$ we have the displacement $(\dot{b}, \dot{c}, \dot{v})$ where $\dot{v} := \frac{1}{2}\mu^{-1/2}\dot{\mu}e$. We thus obtain the appropriate metric from the local norm given by

$$\|(\dot{b}, \dot{c}, \dot{\mu})\|_{(b,c,\mu)} = \mu^{-1/2}(\|\mu^{-1/2}\dot{\mu}e - \dot{q}\|_2^2 + \|\dot{q}\|_2^2)^{1/2}, \tag{4.13}$$

where $\dot{q}$ is defined below (4.11).

Finally, we combine these two cases to consider feasible-interior-point central-path-following methods. Either by replacing $v$ and $\dot{v}$ in (4.12) as in the previous paragraph, or by setting $\dot{q} = 0$ in (4.13), we find that for $\Lambda = \{\mu\} := \mathbb{R}_{++}$, the appropriate metric comes from the local norm

$$\|\dot{\mu}\|_\mu = n^{1/2}\mu^{-1}|\dot{\mu}|. \tag{4.14}$$

Again, the next section gives the corresponding distances and shortest-path geodesics for this case.

We note that only in the feasible cases are the local norms given in closed form in terms of the parameters alone, as in (4.12) and (4.14). In the more general cases, the local norm also involves $\dot{q}$, which is a function of the parameter changes $\dot{b}$ and $\dot{c}$, but uses projections depending on the corresponding solution vector $z$. It is for this reason that determining the corresponding distances and shortest-path geodesics is in general intractable for the infeasible case, and we can only obtain insight from very special cases.

# 5 Shortest paths and examples

This section calculates some shortest-path geodesics for the metrics derived in the previous section.

We start with the simplest case: feasible-interior-point central-path-following methods, where $\Lambda = \{\mu\} := \mathbb{R}_{++}$. In this case the metric is given by the local norm

$$\|\dot{\mu}\|_\mu = n^{1/2}\mu^{-1}|\dot{\mu}|$$

by (4.14). Consider the mapping $\mu \to n^{1/2}\ln\mu$ from $\Lambda$ to $\mathbb{R}$. We see that this is an isometry between $\Lambda$ with the metric given above and $\mathbb{R}$ with the usual Euclidean metric. It follows that the shortest-path geodesic between $\mu_0$ and $\mu_1 < \mu_0$ in $\Lambda$ is just the segment $[\mu_1, \mu_0]$ and, less trivially, that

$$d_\Lambda(\mu_0, \mu_1) = n^{1/2}|\ln(\mu_0/\mu_1)|. \tag{5.1}$$

Thus as long as Hypothesis 1 is true, Theorem 2 shows that

$$O(n^{1/2}|\ln(\mu_0/\mu_1)|) \tag{5.2}$$

iterations are sufficient to move from an approximate center corresponding to the parameter $\mu_0$ to that corresponding to $\mu_1$. This agrees with the complexity bounds in the usual analyses; see Monteiro–Adler [13] and Kojima–Mizuno–Yoshise [9].

Now we turn to feasible target-following methods. Thus $\Lambda = \{v\} := \mathbb{R}^n_{++}$, and the metric is given by

$$\|\dot{v}\|_v = 2n^{1/2}\frac{\|\dot{v}\|_2}{\|v\|_2} \tag{5.3}$$

according to (4.12). Again we seek an isometry to a subset of Euclidean space. Consider the mapping $v \to (\rho, u) := (\ln\|v\|, v/\|v\|_2) \in \mathbb{R} \times S^n_{++}$, where $S^n_{++} := \{u \in \mathbb{R}^n_{++} : \|u\|_2 = 1\}$. We find $v = \exp(\rho)u$, so $\dot{v} = \exp(\rho)[\dot{\rho}u + \dot{u}]$. Since $u^\top u \equiv 1$ implies $u^\top \dot{u} = 0$, we have $\|\dot{v}\|_2 = \exp(\rho)\|(\dot{\rho}, \dot{u})\|_2$, so

$$\|\dot{v}\|_v = 2n^{1/2}\|(\dot{\rho}, \dot{u})\|_2. \tag{5.4}$$

Shortest-path geodesics then correspond in $(\rho, u)$-space to moving at uniform speed from $\rho_0$ to $\rho_1$ and at uniform speed along the great circle from $u_0$ to $u_1$. We find

$$d_\Lambda(v_0, v_1) = 2n^{1/2}\left[\left(\ln\frac{\|v_0\|_2}{\|v_1\|_2}\right)^2 + \left(\arccos\frac{v_0^\top v_1}{\|v_0\|_2\|v_1\|_2}\right)^2\right]^{1/2}. \tag{5.5}$$

For example, the shortest-path geodesic from some $v_0 \in \Lambda$ to a point $v_1$ with $\|v_1\| = \epsilon$ is the straight line segment from $v_0$ to $v_1 := \epsilon v_0/\|v_0\|_2$. This corresponds to following a weighted path; see, e.g., Ding and Li [3].

Let us observe that these geodesics differ from the paths recommended in Jansen–Roos–Terlaky–Vial [5, 6]; their paths always become more centered, while $v_1 = \epsilon v_0/\|v_0\|_2$ implies that ours maintain the same degree of centrality. The main reason for this discrepancy is that Hypothesis 1 does not hold generally in this case. It is necessary to restrict $\Lambda$ to triples $(b, c, v)$ where $v > \theta\mu^{1/2}e = \theta\|v\|_2 n^{-1/2}e$ for some fixed $\theta \in (0, 1]$, and then the hypothesis holds with $\eta$ and $\epsilon$ depending strongly on $\theta$.

Jansen *et al.* define a distance $\delta(v, \bar{v})$ which, for infinitesimally close points, corresponds to the local norm

$$\|\dot{v}\|'_v := \frac{\|\dot{v}\|}{\min(v)}, \tag{5.6}$$

where $\min(v) := \min_j(v^{(j)})$, the smallest component of $v$. The same norm is implicit in the second neighborhood used by Mizuno [11]. This differs by at most a multiplicative constant from $\|\dot{v}\|_v$ in (5.3) as long as $v \geq \theta\mu^{1/2}e$, but the constant depends strongly on $\theta$. Section 2 of [11] and Section 3 of each of [5, 6] show the appropriateness of the measure (5.6); roughly, if $v$ is moved by a small (respectively moderate) distance according to this measure, a single Newton step (bounded number of steps) will yield a good approximation.

The local norm (5.6) 'corresponds' to the local norm in solution space that differs from (3.9) in that $\mu^{-1/2}$ is replaced by $(\min(v))^{-1}$, where $v := x^{1/2}s^{1/2}$. The reason for the quotes above, and the reason we did not use this local norm, is that it is not smooth, but only piecewise smooth, so that we do not obtain Riemannian metrics. However, it is quite possible, based on the results of Jansen *et al.*, that this is a more appropriate, if less smooth, metric for the target-following case. Note that the metrics coincide on the central path, corresponding to choosing $\theta = 1$ so that $\min(v) = \mu^{1/2}$.

Finally, we consider an infeasible case. Here we only address a particular instance.

EXAMPLE Let $A = [I, 0] \in \mathbb{R}^{m \times n}$, and similarly partition the vectors $c = (c_f; c_u)$, $x = (x_f; x_u)$, and $s = (s_f; s_u)$, so that our problems are

$$\text{(LP)} \quad \begin{array}{rrcrcl} \min & c_f^\top x_f & + & c_u^\top x_u & & \\ & x_f & & & = & b, \\ & x_f, & & x_u & \geq & 0, \end{array}$$

and

$$\text{(LD)} \quad \begin{array}{rrcrcl} \max & b^\top y & & & & \\ & y & + & s_f & = & c_f, \\ & & & s_u & = & c_u, \\ & & s_f, & s_u & \geq & 0. \end{array}$$

(For the primal, subscript '$f$' denotes fixed, while '$u$' denotes unconstrained.) We will just consider central-path-following methods, so that our parameter space is $\Lambda = \{(b, c, \mu)\} := \mathbb{R}^m_{++} \times (\mathbb{R}^m \times \mathbb{R}^{n-m}_{++}) \times \mathbb{R}_{++}$. We find, for $\lambda = (b, c, \mu) \in \Lambda$, $z(\lambda) = (x, y, s)$ where

$$x_f = b,\ x_u = \mu c_u^{-1},\ y = c_f - \mu b^{-1},\ s_f = \mu b^{-1},\ s_u = c_u.$$

Thus $H = \mathrm{diag}(\mu b^{-2}, \mu^{-1} c_u^2)$, $J = \mathrm{diag}(\mu b^{-2})$, $Q = \mathrm{diag}(\mu b^{-2}, 0)$, and $P = \mathrm{diag}(0, \mu c_u^{-2})$. Thus

$$\dot{p} = \begin{pmatrix} \dot{c}_f \\ \dot{c}_u \end{pmatrix} - \begin{pmatrix} \dot{c}_f \\ 0 \end{pmatrix} + \begin{pmatrix} \mu b^{-2} \dot{b} \\ 0 \end{pmatrix} = \begin{pmatrix} \mu b^{-2} \dot{b} \\ \dot{c}_u \end{pmatrix}$$

and $\dot{q} = \begin{pmatrix} \mu^{1/2} b^{-1} \dot{b} \\ \mu^{1/2} c_u^{-1} \dot{c}_u \end{pmatrix}$. Now we use (4.13) to obtain

$$\|(\dot{b}, \dot{c}, \dot{\mu})\|_{(b,c,\mu)} = \left( \left\| \begin{pmatrix} \mu^{-1} \dot{\mu} e_f - b^{-1} \dot{b} \\ \mu^{-1} \dot{\mu} e_u - c_u^{-1} \dot{c}_u \end{pmatrix} \right\|_2^2 + \left\| \begin{pmatrix} b^{-1} \dot{b} \\ c_u^{-1} \dot{c}_u \end{pmatrix} \right\|_2^2 \right)^{1/2}.$$

(Note again that this is not strictly a metric, since $c_f$ is not involved, as we should expect; $c_f$ only affects the free variable $y$. It *is* a metric when we restrict to the subvector $(b, c_u, \mu)$.)

Now consider the map $(b, c, \mu) \to (\ln b, \ln c_u, \ln \mu)$ (componentwise). Then our metric above is induced by a fixed ellipsoidal metric in the log-space. Hence shortest-path geodesics in $\Lambda$ correspond to straight lines in $(\ln b, \ln c_u, \ln \mu)$-space, and

$$d_\Lambda((b_0, c_0, \mu_0), (b_1, c_1, \mu_1)) = \left\| \begin{pmatrix} \ln(\mu_0/\mu_1) e_f - \ln(b_0 b_1^{-1}) \\ \ln(\mu_0/\mu_1) e_u - \ln(c_{0u} c_{1u}^{-1}) \\ \ln(b_0 b_1^{-1}) \\ \ln(c_{0u} c_{1u}^{-1}) \end{pmatrix} \right\|_2.$$

We remark that these geodesics are quite different from straight lines in $(b, c, \mu)$-space, as used by most infeasible-interior-point methods. One consequence is that we can obtain better complexity bounds: if

$$2^{-L} e_f \le b_i \le 2^L e_f \text{ and } 2^{-L} e_u \le c_{iu} \le 2^L e_u$$

for $i = 0, 1$, then

$$d_\Lambda((b_0, c_0, \mu_0), (b_1, c_1, \mu_1)) = O(n^{1/2}[\ln(\mu_1/\mu_0) + L]),$$

and the number of iterations of a central-path-following method is of the same order, whereas many infeasible-interior-point methods replace the $n^{1/2}$ with $n^\alpha$ for $\alpha \ge 1$. We hasten to add that this is just one instance, which is trivial to solve directly; but it may suggest the value of trying to obtain geodesics for more general problems.

Finally, consider the extremely trivial case with $m = n = 1$, so that $b$ is one-dimensional and $c_u$ disappears. Then shortest-path geodesics are straight lines in $(\ln b, \ln \mu)$ space. If $b$ moves through a smaller multiplicative range than $\mu$, these geodesics have the property that feasibility is attained at a slower rate than optimality, in strong contradiction to the conventional wisdom that the reverse should be the case.

# 6 Justification of the metric

Here we will discuss Hypothesis 1 for linear programming and also consider two other aspects of our approach: how to recognize good approximations and whether good approximations yield acceptable solutions to the original problems. Almost all proofs will be omitted, and we concentrate on central-path-following methods.

We will use the second condition for $\eta$-approximations, i.e.,

$$\|z - \bar{z}\|_{\bar{z}} \leq \eta \tag{6.1}$$

from (3.4), in confirming Hypothesis 1. This raises the natural question: given only $z \in \mathcal{Z}$ and $\lambda \in \Lambda$, how can we check (6.1) for $\bar{z} = z(\lambda)$? In fact, for the case $\lambda = \mu$ or $\lambda = (b, c, \mu)$ this is straightforward. Suppose $Ax = b$, $A^\top y + s = c$, and $\mu = s^\top x/n$. Then for $\bar{z} = z(\lambda)$ it can be shown that

$$\|\mu^{-1}xs - e\|_2 \leq \delta \leq 0.1 \text{ implies } \begin{cases} \|\bar{x} - x\|_x \leq 3\delta, \ \|\bar{s} - s\|_s \leq 3\delta \\ \|x - \bar{x}\|_{\bar{x}} \leq 9\delta/2, \ \|s - \bar{s}\|_{\bar{s}} \leq 9\delta/2 \\ \|z - \bar{z}\|_{\bar{z}} \leq 7\delta. \end{cases} \tag{6.2}$$

Note that the condition on the left-hand side can easily be checked, and that if it holds for $\delta \leq \min\{\eta/7, 0.1\}$ we know that (6.1) holds.

Next we address whether having a good approximation $z$ to an acceptable solution $\bar{z}$ suffices. Since we are using a Newton step at each iteration, linear constraints will be satisfied exactly, so that if $A\bar{x} = b_1$ and $A^\top\bar{y} + \bar{s} = c_1$, the same will be true for $x$ and $(y, s)$. Also, (6.1) will ensure that $x > 0$ and $s > 0$ for sufficiently small $\eta$, so feasibility is assured. The only remaining concern is the duality gap $s^\top x$. We find

$$\begin{aligned} s^\top x &= [\bar{s} + (s - \bar{s})]^\top[\bar{x} + (x - \bar{x})] \\ &\leq \bar{s}^\top\bar{x} + \|s - \bar{s}\|^*_{\bar{w}}\|\bar{x}\|_{\bar{w}} + \|\bar{s}\|^*_{\bar{w}}\|x - \bar{x}\|_{\bar{w}} + \|s - \bar{s}\|^*_{\bar{w}}\|x - \bar{x}\|_{\bar{w}}, \end{aligned}$$

where $\bar{w} := \bar{x}^{1/2}\bar{s}^{-1/2}$. But (6.1) ensures that $\|x - \bar{x}\|_{\bar{w}}$ and $\|s - \bar{s}\|^*_{\bar{w}}$ are at most $\mu^{1/2}\eta$, where $\mu := \bar{s}^\top\bar{x}/n$, and it is not hard to show that $\|\bar{x}\|_{\bar{w}} = \|\bar{s}\|^*_{\bar{w}} = \mu^{1/2}n^{1/2}$, so that

$$s^\top x \leq 1.3\bar{s}^\top\bar{x} \tag{6.3}$$

as long as $\eta \leq 0.1$.

The fundamental reason that Hypothesis 1 holds is the quadratic convergence of Newton's method, but we need explicit constants and also many applications of the fact that norms evaluated at neighboring points are close. The basic property here is that

$$\|x_+ - x\|_x \leq \delta < 1 \text{ implies } \|v\|_{x_+} \leq (1-\delta)^{-1}\|v\|_x \tag{6.4}$$

for any $x, x_+ \in \mathbb{R}^n_{++}$ and $v \in \mathbb{R}^n$. This is trivial to show directly, since

$$\|v\|_{x_+} = \|(X_+^{-1}X)(X^{-1}v)\|_2 \leq \|X_+^{-1}X\|_2\|X^{-1}v\|_2.$$

Note that from (6.4) the second line of implications in (6.2) follows from the first, since $(1-3\delta)^{-1}3\delta \leq (0.7)^{-1}3\delta \leq 9\delta/2$. (The third line then follows directly from the definition (3.11).)

Finally, let us state two results which are key ingredients in establishing Hypothesis 1.

**Lemma 4** *Let $x, x_+ \in \mathbb{R}^n_{++}$ with $\delta := \|x_+ - x\|_x < 1$. Then*

$$\| - x_+^{-1} + x^{-1} - X^{-2}(x_+ - x)\|_x^* \leq \frac{\delta^2}{1-\delta}. \tag{6.5}$$

Note that the norm on the left-hand side can also be written in terms of the logarithmic barrier function $\phi$ defined in (3.10): it becomes

$$\|\phi'(x_+) - \phi'(x) - \phi''(x)(x_+ - x)\|_x^*.$$

This makes it clear that the lemma is bounding the error in the first-order Taylor approximation to $\phi'(x_+)$.

The next result refers to the points appearing in Hypothesis 1, and assumes a central-path-following method, so that $\lambda = \mu$ or $\lambda = (b, c, \mu)$.

**Lemma 5** *If $z, \bar{z}_+$, and $z_+$ are as in Hypothesis 1, with $\bar{z}_+ = z(\mu_+)$ or $z(b, c, \mu_+)$, and $H = W^{-2}$ with $w = x^{1/2}s^{-1/2}$, then*

$$\|x_+ - \bar{x}_+\|_w^2 + \|s_+ - \bar{s}_+\|_w^{*2} = \|\mu_+(-\bar{x}_+^{-1} + x^{-1}) - H(\bar{x}_+ - x)\|_w^{*2}. \tag{6.6}$$

Note that the left-hand side of (6.6) is closely related to $\|z_+ - \bar{z}_+\|_{\bar{z}_+}$ (except that the norm is wrong), while its right-hand side is closely related to the quantity bounded in (6.5) (except that the norm is wrong and $H/\mu_+$ replaces $X^{-2}$). Putting all these pieces together enables one to prove

**Theorem 6** *For (feasible or infeasible) central-path-following methods (with $\lambda = \mu$ or $\lambda = (b, c, \mu)$), Hypothesis 1 holds with $\epsilon = \eta = 0.04$.*

We briefly mention, as we have hinted above, that in the case of target-following methods it is necessary to restrict $v$ so that $v \geq \theta\|v\|_2 n^{-1/2}e$, where $\theta \in (0, 1]$ is a constant. Thus $\Lambda = \{v\} := \{v \in \mathbb{R}^n_{++} : v \geq \theta\|v\|_2 n^{-1/2}e\}$ or $\Lambda = \{(b, c, v)\} := A(\mathbb{R}^n_{++}) \times (A^T(\mathbb{R}^m) + \mathbb{R}^n_{++}) \times \{v \in \mathbb{R}^n_{++} : v \geq \theta\|v\|n^{-1/2}e\}$. With this restriction, Hypothesis 1 holds with $\epsilon = \eta = 0.04\theta$; note the unpleasant dependence on $\theta$.

# References


[1] Boothby, W.M. (1986) *An Introduction to Differentiable Manifolds and Riemannian Geometry*, Academic Press (New York).

[2] Davidon, W.C. (1991) "Variable metric methods for minimization", Report ANL-5990, Argonne National Laboratories, Argonne, IL, 1959, reprinted in *SIAM J. Optimization* **1**, 1–17.

[3] Ding, J. and T.Y. Li. (1990) "An algorithm based on weighted logarithmic barrier functions for linear complementarity problems", *Arabian J. Science Engng*, **15**(4), 769–685.

[4] Fletcher, R. and M.J.D. Powell (1963) "A rapidly convergent descent method for minimization", *Computer J.* **6**, 163–168.

[5] Jansen, B., C. Roos, T. Terlaky and J.P. Vial (1993) "Primal-dual target-following algorithms for linear programming", Technical Report 93–107, Faculty of Technical Mathematics and Informatics, TU Delft, NL-2600 GA Delft, The Netherlands.

[6] Jansen, B., C. Roos, T. Terlaky and J.P. Vial (1996) "Long-step primal-dual target-following algorithms for linear programming", *Zeitschrift für Operations Research – Mathematical Methods of Operations Research* **44**(1).

[7] Karmarkar, N.K. (1984) "A new polynomial-time algorithm for linear programming", *Combinatorica* **4**, 373–395.

[8] Kojima, M., S. Mizuno and A. Yoshise (1989) "A primal-dual interior point algorithm for linear programming", in *Progress in Mathematical Programming: Interior Point and Related Methods* (N. Megiddo, ed.), Springer Verlag (New York), 29–47.

[9] Kojima, M., S. Mizuno and A. Yoshise (1989) "A polynomial-time algorithm for a class of linear complementarity problems", *Mathematical Programming* **44**, 1–26.

[10] Lemke, C.E. (1965) "Bimatrix equilibrium points and mathematical programming", *Management Science* **11**, 681–689.

[11] Mizuno, S. (1992) "A new polynomial time method for a linear complementarity problem", *Mathematical Programming* **56**, 31–43.

[12] Mizuno, S., M.J. Todd and Y. Ye (1995) "A surface of analytic centers and primal-dual infeasible-interior-point algorithms for linear programming", *Maths Operations Research* **20**, 135–162.

[13] Monteiro, R.D.C. and I. Adler (1989) "Interior path following primal-dual algorithms: Part I: Linear programming", *Mathematical Programming* **44**, 27–41.

[14] Nesterov, Yu.E. and A.S. Nemirovskii (1993) *Interior Point Polynomial Methods in Convex Programming: Theory and Algorithms*, SIAM Publications. SIAM (Philadelphia).

[15] Nesterov, Yu.E. and A.S. Nemirovskii (1995) "Multi-parameter surfaces of analytic centers and long-step surface-following interior point methods", Research Report 3/95, Faculty of Industrial Engineering and Management, Technion, Haifa 32000, Israel.

[16] Nesterov, Yu.E. and M.J. Todd (1994) "Self-scaled barriers and interior-point methods for convex programming", Technical Report No. 1091, School of Operations Research and Industrial Engineering, Cornell University. To appear in *Mathematics of Operations Research.*

[17] Nesterov, Yu.E. and M.J. Todd (1995) "Primal-dual interior-point methods for self-scaled cones", Technical Report No. 1125, School of Operations Research and Industrial Engineering, Cornell University.

[18] Polyak, R. (1992) "Modified barrier functions (theory and methods)", *Mathematical Programming* **54**, 177–222.

[19] Powell, M.J.D. (1992) "The complexity of Karmarkar's algorithm for linear programming", in *Numerical Analysis 1991* ( D.F. Griffiths and G.A. Watson, eds), volume 260 of *Pitman Research Notes in Mathematics*, Longman (Burnt Hill, UK), 142–163.

[20] Powell, M.J.D. (1993) "On the number of iterations of Karmarkar's algorithm for linear programming", *Mathematical Programming* **62**, 153–197.

[21] Powell, M.J.D. (1995) "Some convergence properties of the modified log barrier method for linear programming", *SIAM J. Optimization* **5**, 695–739.